Synthesis Lectures on Engineering, Science, and Technology

The focus of this series is general topics, and applications about, and for, engineers and scientists on a wide array of applications, methods and advances. Most titles cover subjects such as professional development, education, and study skills, as well as basic introductory undergraduate material and other topics appropriate for a broader and less technical audience.

Ferdinand Rivera

STEM Learning in Extended Reality

Research, Design, and Examples

Springer

Ferdinand Rivera
Ed.D. Educational Leadership Program
Department of Mathematics and Statistics
San Jose State University
San Jose, CA, USA

ISSN 2690-0300 ISSN 2690-0327 (electronic)
Synthesis Lectures on Engineering, Science, and Technology
ISBN 978-3-031-78972-4 ISBN 978-3-031-78973-1 (eBook)
https://doi.org/10.1007/978-3-031-78973-1

This Springer imprint is published by the registered company Springer Nature Switzerland AG
The registered company address is: Gewerbestrasse 11, 6330 Cham, Switzerland

Sensorimotor life is, for the most part,
for most people, and for most of the
time, the ongoing bustle of
animate embodied being.

(Di Paolo, Buhrmann, & Barandiaran, 2017, p. 4)

[T]he nature of our bodies, as they interact
with our structured environments, shapes
what things and events are meaningful
to us and how we make sense of
and reason about them.

(Johnson, 2017, p. 221)

Preface

Science, technology, engineering, and mathematics (STEM) education is in a time of great ferment—that is, an exhilarating moment in the history of the present characterized by incitation, stirring up, trouble, disorder, agitation, and excitement—caused from the outside by remarkable developments in extended reality (XR), which could very well be "the learning aid of the 21st century" (Rogers, 2019). XR stands for all immersive digital tools, apps, and learning environments, from the simple structure of augmented reality (AR) that is grounded in some fixed physical space, to the imagined and fully manipulable space of virtual reality (VR), to a mixed reality (MR) world where both real and digital elements mutually interact to support wanderings and produce knowledge of one's own making. Consistent with the publisher's intent, this book is short and focused, and is meant to serve as a bridge between a journal article and an average book. In four chaps., I explore ideas in 4E cognition—embodied, embedded, extended, and enactive—that could serve as a useful theoretical framework for thinking with/in XR (Chap. 1), discuss some recent research results in XR and STEM education (Chap. 2), describe how XR is changing the nature of learning and assessment (Chap. 2), propose a design framework for modeling deep learning in XR-mediated activity (Chap. 2), and provide examples of XR apps from the STEM disciplines (Chaps. 3 and 4). I note at the outset that this book does not address collaborative XR (metaverses, virtual worlds, etc.) and issues that pertain to inclusive immersion (see Dudley et al., 2023, for an impressive scoping review).

After writing this book and having successfully developed two VR apps for learning fundamental concepts in mathematics, I don't think I can teach my students to learn mathematics the same way ever again (i.e., as disembodied knowers). At the very least, XR has irremediably reshaped my views on engagement and inquiry. I am also reminded of the following words from the digital age visionary, Marshall McLuhan, which capture how I felt when I started exploring XR as a learning tool: "Anybody moving into a new world loses identity" (quoted in Benedetti & DeHart, 1997, p. 100). The new identity that results from authentically engaging in XR is onto-epistemological in Karen Barad's sense;

while we own the knowledge we construct, the authoring process is itself dependent on our bodily intra-actions with XR, which we endow with the agentic ability to influence and transform not only what and how we think, but also who and where we are in either a hybrid or fully digital environment. There is, in other words, a relationship of "mutual entailment," one in which our emerging knowledge and XR are "mutually articulated" in a space that is "neither ontologically nor epistemologically prior" (Barad, 2007, p. 820). Certainly, more intentional collaboration with others will likely cause our constructed knowledge to evolve, mature, deepen, and emerge to be as good as or even better than institutional knowledge. Finally, from what I could infer from the rich relevant research literature on XR and STEM education, well-conceptualized XR apps in educational settings, and the learning environments they support, could be the needed humdinger to reboot a traditional curriculum, transforming it into something that is living, breathing, and exciting to learn once more.

Thanks to the National Science Foundation and San Jose State University's Research, Scholarship and Creative Activity Program for supporting innovative research. All the ideas in this book are mine and do not necessarily reflect the views of the foundation and my university.

San Jose, CA, USA Ferdinand Rivera

Contents

1 Cognition Meets Extended Reality

Current extended reality applications ("apps"), and the learning environments they support, provide users with the bodily experience of manipulating real and fictional objects, and consequently, creating and establishing knowledge based on their actions and interactions with those objects. *Extended reality* (XR) refers to all types of digitally-mediated immersive technologies that endow users with the ability to explore and construct ideas in highly interactive spaces that extend the limits of the physical models and representations being enhanced, simulated, or replaced. Well-designed XR environments also allow users to "see the unseen" (Clark, 2008; Faulconer & Gruss, 2018; Potkonjak et al., 2016) and provide the "feeling of being there" (Ferrara et al., 2021), which cannot be conveniently performed with other non-XR apps. Some of the XR environments are, in fact, able to move beyond the constraints of what is possible to accomplish in real environments (Slater & Sanchez-Vives, 2016). Keen's (2018) following words about XR certainly resonate with users who find XR apps valuable in learning contexts: They "not only empower us, like cinematic superheroes, to walk on Mars but also to experience what it feels like to be a cow or a tree" (p. 1). The hope is that the heightened, emotional, personal, "extremely motivating" (Kaufmann, 2009), and direct experience in this form of "edutainment" within such "playful learning environments" (Kiourt et al., 2020) may have a positive and transformative impact on users' learning experiences, behaviors, and actions (Massaro, 2019, p. 497) and "in ways [they] will not forget for the rest of their lives" (Kaufmann, 2009, p. 149). More importantly, from a science, technology, engineering, and mathematical (STEM) perspective, meaningful impact of XR apps is related to the development of a deep understanding of, and appreciation for, *structures* and growth in *structure sense.* That is, interesting and purposeful real- and digitally-constructed objects and spaces model structures that consist of well-defined elements in either a known or an

F. Rivera, *STEM Learning in Extended Reality*, Synthesis Lectures on Engineering, Science, and Technology, https://doi.org/10.1007/978-3-031-78973-1_1

emergent system with clear and explicit rules for manipulating those elements. Of course, it is worth noting that some non-STEM XR apps, such as those that are currently being used to explore heritage learning in areas like architecture, history, archeology, and music, provide virtual experiences that attend to the affective domain of learning, for example social and cultural sensitivity (e.g., human dignity, empathy, and respect; Schaper & Pares, 2022; see Innocente et al.'s [2023] meta-synthesis of 65 studies that describe characteristics of effective XR systems in this field). Also, like all other STEM XR apps, they are meant to assist users in those fields attain knowledge and increase their motivation (Dordio et al., 2024).

Kaufmann (2009) points out the inherently three-dimensional nature of all XR environments, which allow learners to fully interact with digital objects that they can virtually touch and feel with either their hands or some other external device (e.g., a touch controller or a pen). For instance, Fig. 1.1 captures a situation in the virtual reality *Transforming Graphs* app that immerses beginning algebra students in a virtual classroom. In this fully immersive environment, they can grab the graph of a function with either hand and move it according to the mathematical transformation they are exploring. Graphing a function on a Cartesian plane with Desmos, a popular two-dimensional interactive graphical software, already provides a learning environment that allows them to manipulate graphs of functions. However, in the virtual reality environment, as illustrated in Fig. 1.1, the equation $y = 1.0(x - 4.3) + 7.1$ emerges as a recording of two bodily movements that can take place in any order: pinching the right (or left) hand and motioning it to the right moves the graph of $y = x$ horizontally 4.3 units to the right, and motioning the right (or left) pinched hand upward translates the new graph 7.1 units vertically upward. This particular type of physical body-virtual environment conceptual coupling far exceeds the educational advantages of traditional patty paper or Desmos process that we sometimes see happen in today's beginning algebra classes. The virtual environment depends on hand movements that allow students to feel mathematics through their body. Well-designed XR environments in STEM should be able to make this body, space, visualizing, sensing, thinking, and mathematizing happen all at the same time, capturing in an authentic manner what it means to reify concepts—that is, to make concepts alive in 3D. Exceptional ones especially have the power to dissolve any artificial separation or gap between abstract concepts and the concrete models they represent.

XR apps range from simply blending real and digital objects to constructing entirely new virtual spaces based on some predefined structure. Tremosa's (2022) categories of XR shown in Fig. 1.2 describe in simple terms three types of XR apps based on interactions between actual and digital objects. In *augmented reality* (AR), constructed digital objects interact with fixed objects in an actual physical environment (e.g., placing a QR code on a piece of paper or a digital bird on a picture of a tree). In *mixed reality* (MR), both actual and digital objects interact with one another (e.g., transporting learners to impossible places such as the moon or simulated driving under challenging digitally-mediated conditions). MR apps influence our "perception of an experience" because they draw on

(a) Horizontal Translation in the VR Transformations Graph App (Rivera & Paidi, 2024)

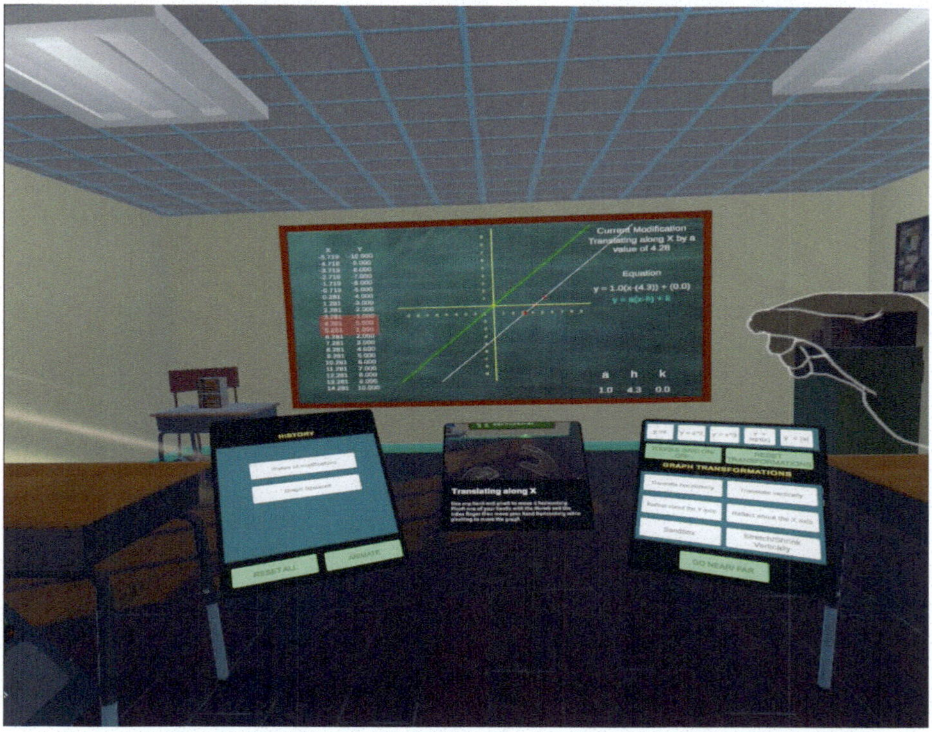

Fig. 1.1 **a** Horizontal translation in the VR transformations graph app (Rivera & Paidi, 2024). **b** Vertical translation in the VR transformations graph app (Rivera & Paidi, 2024)

"motion sensors, body tracking, and eye tracking interplay" to capture some context we are exploring based on what we try to access ourselves (Ziker et al., 2021, p. 56). In *virtual reality* (VR), all objects are digital that can be manipulated in a fully immersive virtual environment (e.g., horizontally or vertically translating graphs of functions in a Cartesian plane such as the one shown in Fig. 1.1). VR apps by design are not dependent on any physical model of an environment.

VR apps could be *mobile* (e.g., Google Cardboard), *head-mounted* (e.g., Meta 2), and *enhanced* (e.g., head mounted with bodysuits) (Khaddage et al., 2023). More recently, too, there are emerging VR systems that extend the usual individual solo experience by allowing multiple users to simultaneously collaborate (i.e., participating in virtual conferences as avatars that support interactions with other participants or *ShapesXR*'s app that allows teams to virtually collaborate on a project in real time). These collaborative immersive virtual environments are often referred to as *virtual worlds* (Ziker et al., 2021). A good example is *CyberMath,* a collaborative VR space that provides students and teachers

(b) Vertical Translation in the VR Transformations Graph App (Rivera & Paidi, 2024)

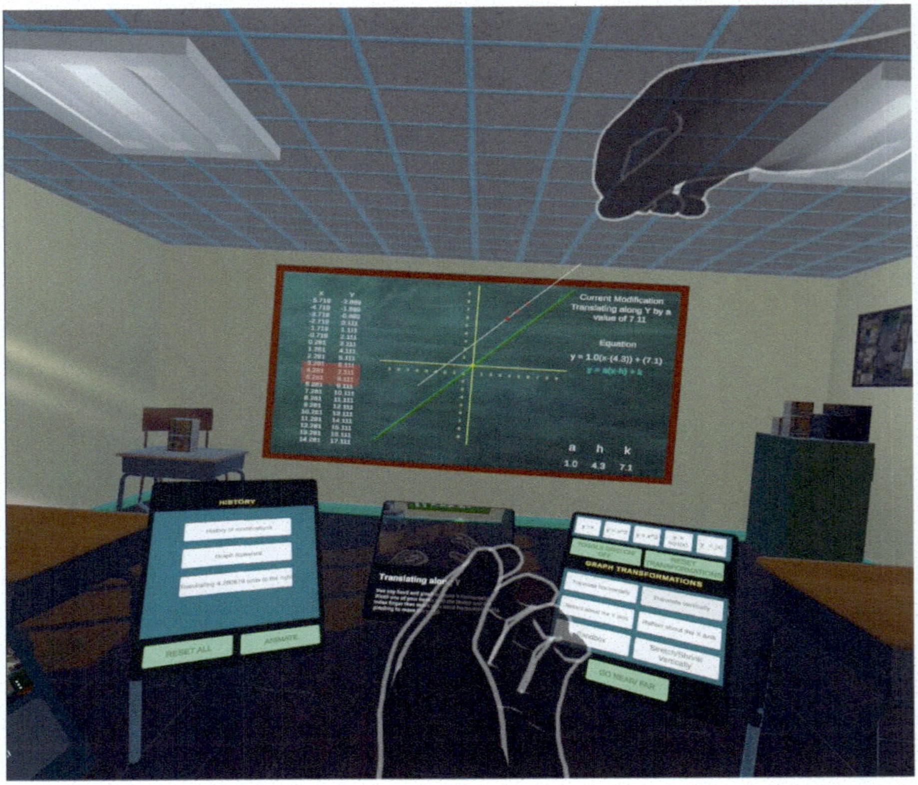

Fig. 1.1 (continued)

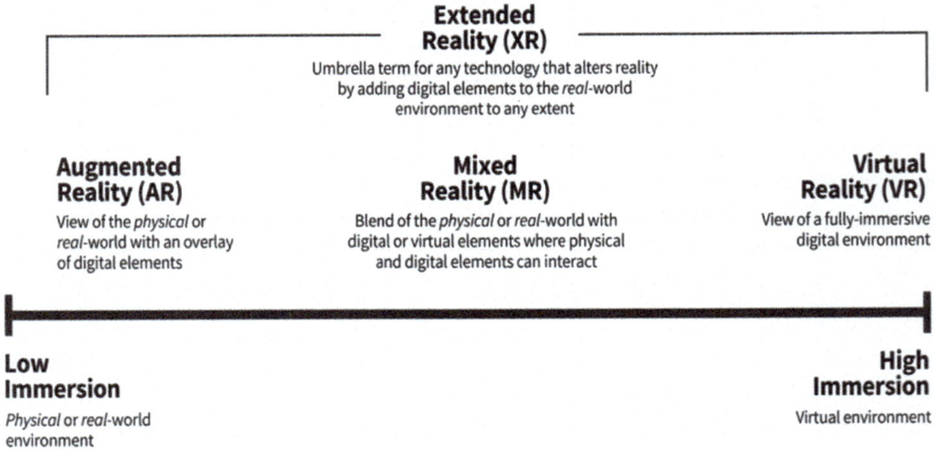

Fig. 1.2 XR categories (Tremosa, 2022)

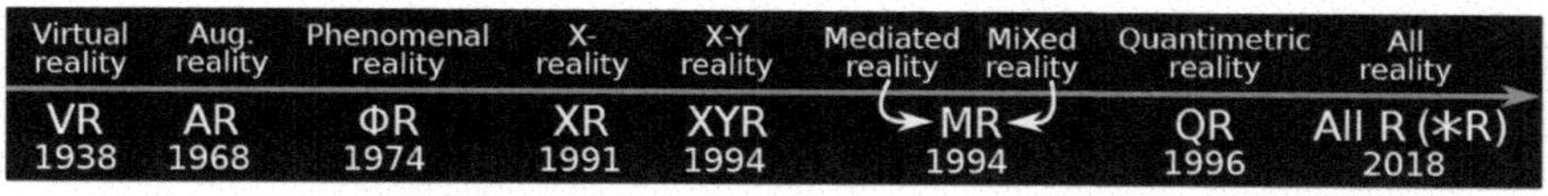

Fig. 1.3 Mann et al.'s (2018) realities timeline

with an opportunity to explore and manipulate abstract mathematical objects together, co-construct knowledge, and develop shared perceptions of objects and concepts. Certainly, these opportunities happen in everyday classrooms and with traditional and classical tools for representing objects and concepts, but it is not often guaranteed.

Mann et al. (2018) developed a more elaborate typology of different realities and the year in which they initially emerged (see Fig. 1.3). The authors' paper offers an interesting read on the evolution of such technologies, especially the various kinds of reality (i.e., synthetic, phenomenal, natural, etc.) that are available to us but tend to take for granted.

Figure 1.4 provides another depiction of the *VR Transforming Graphs* app shown in Fig. 1.1. The app runs on a headset (e.g., Quest) and provides users with both hand grabbing and touch controller options. Beginning algebra students who use it will experience what it means to transform graphs of functions using hand gestures. For instance, pinching both hands and moving them away from each other along the y-axis stretches the graph vertically. The *VR Transforming Graphs* app, like all other VR apps, transports individual learners to a "different time and place" (Ziker et al., 2021) that permits them to engage in inquiry and discovery at their own pace (e.g., what transformational actions produce the same reflection of a graph about the y-axis?).

Figure 1.5 is an example of an AR app that activates 3D images corresponding to the drawn pictures on two geometry textbook pages. An AR's hybrid system could be used with a smartphone or smart pad, but it does not provide the user with the option of fully manipulating either physical or digital elements in an interactive manner. Also, like other AR apps, its current functionalities do not usually include real-time feedback (Ziker et al., 2021).

AR apps are classified by the type of marker or trigger that is used to overlay the digital content on a fixed physical surface. The simple ones, called *Web-based AR,* only require an active hyperlinked URL on, say, a fixed space that when clicked with a camera activates some digital information drawn from the Web. Web-based AR makes the presentation of 2D content more interesting, interactive, and engaging for users. The more complicated ones—*marker-based AR*—use markers, images, or triggers as "reference points for superimposing digital content onto the physical environment" (PlugXR, 2024, n.p.). When a camera recognizes a marker, it "overlays virtual elements, such as 3D models, animations, or information, onto the corresponding marker locations" (ibid.). Figures 1.5 and 1.6 provide examples of sheets of paper with a QR code as a marker, and Fig. 1.7 activates a 3D image of a robot with the yellow square symbol as a marker. *Location-based AR* apps do not necessarily require markers but depend on a real-time "GPS, accelerometer, digital

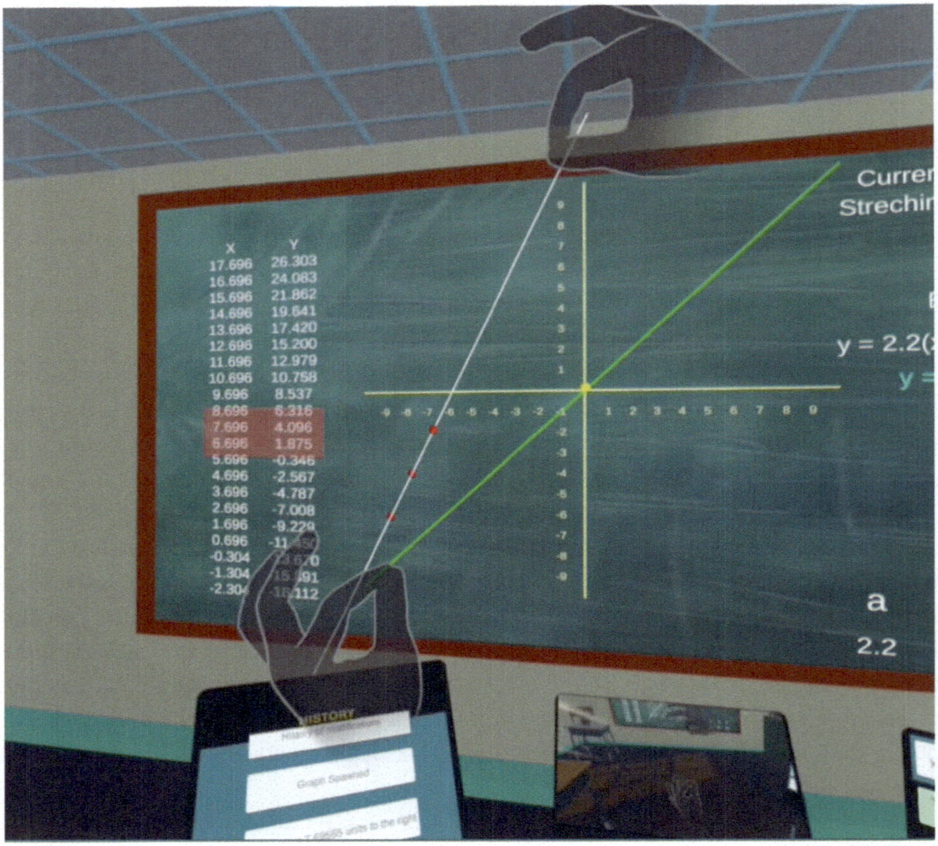

Fig. 1.4 Vertical stretching in the VR transforming graphs app (Rivera & Paidi, 2024)

compass and other technologies to identify a device's location and position with a high level of accuracy" (Altynpara & Chabanovska, 2023, n.p.). To use them, users need to be in an actual location first that would then provide the context for activating the relevant digital information (e.g., 3D graphs and pictures, links, animations, and text) that has been programmed for that location (Wikitude, 2024). Figure 1.8 shows an example of a location-based AR that activates important information about the largest mosque in Kota Kinabalu. Pokemon Go is another good example of a location-based AR. Users typically need to walk around some specific physical location in order to grab Pokémon characters using Poké Balls.

Simeone (2023) offers another way to think about XR—that is, in terms of *presence*, "the subjective experience of being in one place or environment even when one is physically situated in another environment" and "involves feeling physically surrounded by a mediated, but seemingly natural and believable space" (p. 6). Presence is low in AR

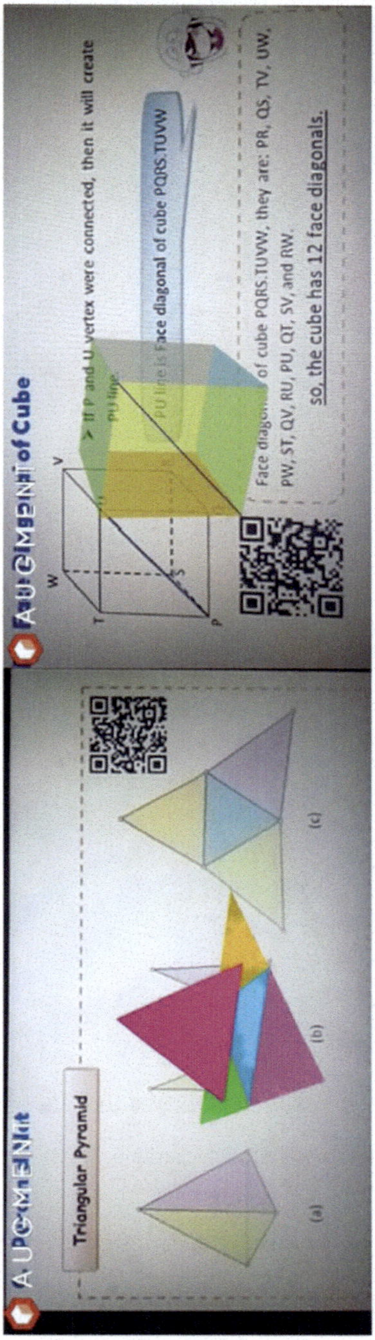

Fig. 1.5 AR geometry (Auliya & Munasiah, 2019)

Fig. 1.6 Example of a QR code marker-based AR (PlugXR, 2024)

Fig. 1.7 Example of an image marker-based AR (Mecharithm, 2022)

and high in VR simply because AR basically adds interactive digital elements in some fixed physical environment with the goal of enhancing user experience. VR, on the other hand, sets up a digital environment that supports "increased user engagement, emotional involvement, and a stronger connection with the VR environment" (Interactive Design Foundation, 2016, n.p.).

Fig. 1.8 Example of a locator-based AR (Mecharithm, 2021)

In this book, we explore key concepts in embodied, enactive, embedded, and extended cognition to help us think about how to use XR apps for thinking and learning in STEM education contexts. Recent XR apps in the STEM disciplines have produced amazing products that would be interesting to evaluate in different ways. My goal is to encourage readers, practitioners, and scholars to develop a theoretical mindset about why and how we want to use XR in STEM. This goal applies to all emergent technologies for that matter. More importantly is the need to assess how we can use these new generation technologies to reconceptualize and advance conversations and research on learning and learning environments in STEM. It is critical to explore how we can change the nature of students' experiences in the classroom and how we can develop more appropriate ways of conceptualizing learning, teaching, and assessment that go beyond business-as-usual. Ziker et al.'s (2021) following point about XR captures in a succinct manner the potentially transformative power of such apps in learning: "We can now consider not just adapting existing educational content, but conceiving of entirely new ways of educating that leverage an embodied experience and allow for a huge range of augmentation" (p. 66).

1.1 The 4E Cognitive Scheme: Embodied, Enactive, Extended, and Embedded Cognition

Gallagher and colleagues (cf. Gallagher, 2023; Newen et al., 2018) use the 4E scheme to explain Embodied Cognition (EC)—that is, embodied, enactive, extended, and embedded (also referred to as wide cognition [Miłkowski et al., 2018]). There are other factors that influence thinking, too, as Gallagher (2023) and other colleagues (e.g., Johnson, 2017) have pointed out, such as ecological, emotional, empathy, existential, evolutionary, and exaptative. However, in this book, we will investigate the extent to which the 4E scheme can be used as a theoretical framework for understanding XR thinking and learning in the STEM fields. At the very least, the 4E scheme shifts away from individualistic views of cognition and toward the view that recognizes the simultaneous roles played by "material embodiment, environmental embedding," interactions, and artifact or tool use in the "molding of cognitive processes" (Smart & Sycara, 2015, p. 3837). Individuals that engage in XR-mediated activity ground their thinking and learning in "sensorimotor activity and are shaped by and reshaped within those environments" (ibid., p. 2). In this chapter, we describe the 4E scheme in detail and reflect on how the fundamental design of almost all XR apps seem to align with the 4E scheme. Also, all the examples we use in this chapter and throughout the book will be confined to XR apps in STEM. Hence, readers who specialize in the use of manipulatives, graphing calculators, and other current interactive apps (e.g., Desmos, Geogebra, simulations) to support learning in the STEM fields will need to establish connections between XR and those tools on their own.

What is Embodied Cognition all about? EC underscores the importance of, and mutually determining relationship, between bodily/organic/neural/extraneural and environmental processes in shaping how and what we think. For instance, a beginning algebra student's mathematical perception of a vertical translation involving the graph of the quadratic function $y = x^2$ shown in Fig. 1.9 using the *VR Transforming Graphs* app emerges from their intentional epistemic action of grabbing the graph in the virtual environment and moving it up and down on the Cartesian plane to see what changes and what stays the same in the corresponding equation. The combination of the student's inductive or repeated actions on the graph, and what they are able to see and touch virtually, enable them to abduce the meaning of the equation $y = x^2 + k$. Consequently, their virtually-mediated generalizations resulting from this combination of abductive and inductive actions on the VR app naturally emerge from what they see, experience, and manipulate directly in the environment.

In Fig. 1.10, students use the VR app to investigate two different ways of obtaining the reflection of the graph of $y = x^3$ about the y-axis. They may position their right (or left) palm vertically and then rotate their palm by 180^0 to produce the graph $y = -x^3$ (Fig. 1.10a). They may also reflect the graph of $y = x^3$ about the x-axis first (by positioning their right (or left) palm horizontally and rotating their palm by 180^0) and then rotate the graph by 180^0 (with the left pinched hand fixed and the right pinched

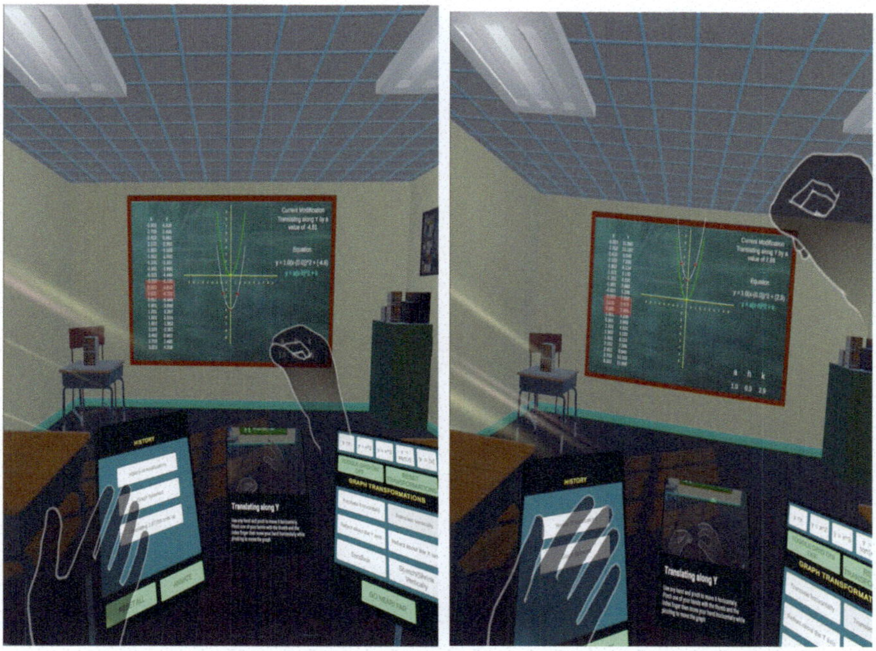

Fig. 1.9 Learning about $y = x^2 + k$ with the VR transforming graphs app (Rivera & Paidi, 2024)

hand performing a 180^0 counterclockwise otation) (Fig. 1.10b). Hence, mathematical perception and thinking and, more generally, perceptions and thinking in XR, are typically emergent and actively constructed depending on what individual learners choose to see, how they act, and what they manipulate and infer based on their sensorimotor experiences in such environments. This actor- and action-oriented view is different from traditional notions that see perception to be a mere byproduct of what could be observed passively through the senses. This particular description of a student's learning in the *VR Transforming Graphs* app also illustrates why EC is conceptually grounded in fields such as phenomenology, pragmatism, robotics, and neuroscience, which counter classic Cartesian philosophy and cognitivist theories of the human mind that formulate cognition solely in terms of some innate computation or information processor that is governed by symbolic rules and logical relationships (Gallagher, 2023).

The 4E cognition scheme— that is, embodied, enactive, extended, and embedded— offers a unified approach in understanding EC. Interested folks are strongly encouraged to read Gallagher (2023) and related resources (e.g., Menary, 2010; Newen et al., 2018) for a deeper dive into its history, including points of convergence and divergence among the four approaches. The remaining sections below draw quite heavily on Gallagher (2023), but only those aspects that seem relevant and sufficient enough to deepen our understanding of thinking, learning, and living within XR spaces are considered here.

(a) Reflect about the y axis in one step

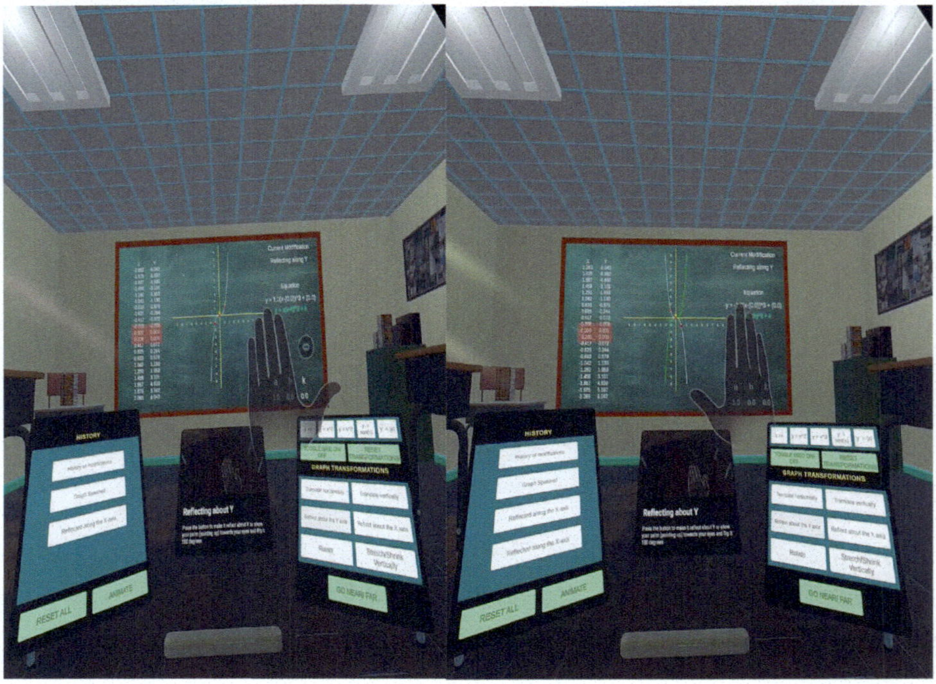

Fig. 1.10 **a** Using the VR transforming graphs app to obtain the reflection of $y = x^3$ about the Y-axis

1.2 Embodied Component in EC

Embodiment is either strong or weak. In *strong embodiment,* all our cognitive processes are endowed with physiological and sensorimotor properties; it assumes that cognitive and physiological processes are necessarily linked, which then allow us to see the value and central role of body-mediated experience. In *weak embodiment,* the body is not as important since the brain is seen to be central to cognitive processes, which explains the "mind-as-a-computer" or "body-in-the-mind" metaphors. Weak embodiment shares many of the same views as classical cognitivist accounts and suggests with some good evidence the existence of embodied simulations or body-formatted and body-related mental representations (Goldman & de Vignemont, 2009) that emerge from neural circuits in the brain, which then enable individuals to imagine, say, translating an oblique line in a Cartesian plane vertically via its y-intercept without needing to physically demonstrate it in, say, XR. In other words, acting on those objects may take place in the brain alone. The current dilemma with strong embodiment is the lack of strong empirical evidence linking the body or sensorimotor actions and higher cognitive processing, which may indicate that

(b) Reflect about the y axis in two steps

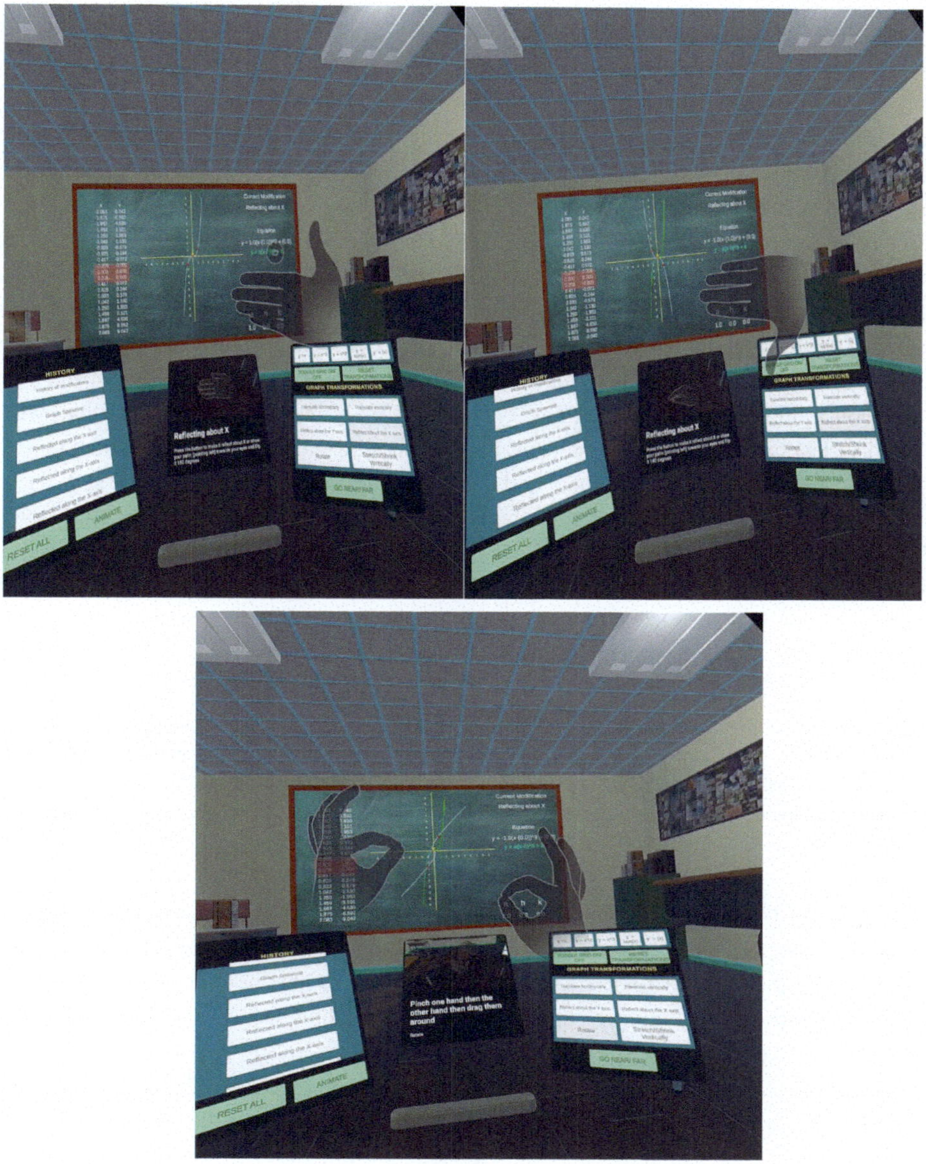

Fig. 1.10 (continued)

cognition may be too complicated to explain via strong embodiment alone (Tirado et al., 2018).

The succeeding chapters in this book will provide different kinds of evidence that convey the strength of strong embodiment in XR activity for STEM learners. Whether immersion in XR activity is low, high, or in between, XR makes the body the necessary and central tool for constructing and structuring knowledge through action. Gallagher (2023) mentions the work of Lakoff and Johnson (1991), who claim that the "structure, composition, and motor abilities of the body determine how we experience things" and "also determine what we experience, and how we understand the world" (p. 9). Furthermore, the two authors point out that "our conceptual life begins in spatial and motor behaviors and derives meaning from bodily experience. Accordingly, the 'peculiar nature of our bodies shapes our very possibilities for conceptualization and categorization' (Lakoff & Johnson, 1999, p. 19)" (ibid.). This description of the essential role of the body in the conceptualization process aligns with the XR body-mediated experience that, when carefully designed and implemented, provides learners with the necessary "metaphors" for "bridg[ing] embodied experience and conceptual thought" (Gallagher, 2023, p. 9).

Drawing on existing evidence, Gallagher (2023) underscores two additional points about body-environment interactions in strong embodiment: First, the way our body is positioned and moves affect what we perceive and conceptualize, and that our movements are a consequence of some kind of *soft assembly* process, a term that Clark (2007) describes in the following manner below.

> Movements can be seen as 'softly-assembled' patterns created and dissolved as tasks and environments change, with some patterns easy and preferred, and others more difficult and unstable.... Moreover, as these synergies are assembled, they also take advantage of the non-neural aspects of movement: effects of gravity, elastic properties of muscles and inertial effects (Thelen & Bates, 2003, p. 381).

Also, many of those self-assembled solutions that emerged to meet certain demands may be deeply "integrated, creating whole new unified systems of distributed problem-solving" (Clark, 2007, p. 103). Second, affective factors such as our emotions, mood, and other physical sensations (e.g., hunger, fatigue) limit what and how we are able to perceive, act, and cognitively process.

The above two points are particularly important in any account of XR design, learning, and activity. At the very least, XR environments that are not flexible and sensitive enough to meet user demands and characteristics are not likely going to support meaningful interactions and self-assembled strategies. Both the user's movement and the XR environment need to co-evolve in the same way the brain and the body co-evolve in a productive manner to produce meaning. Also, passive learning or unstructured actions that are not aligned with the goals of the XR activity will most likely limit what students are able to construct. I remember a few frustrating experiences of unproductive thinking and acting (e.g., unstructured tweaking) involving several beginning linear algebra students who tested the *VR Vector Manipulations* app shown in Fig. 1.11. One student simply kept clicking on the different tool boxes (white boxes) to solve a given task. He thought that his approach, which he acquired from playing VR games, would help him solve the task quickly. Another student was impatient about learning to use the touch controllers effectively, which made it difficult for her to move the vectors to meet certain conditions. A third student merely relied on my directions for moving vectors around the space but could not infer anything deeper than what she saw (e.g., when she moved all three vectors so that they coincided to form a single line that passed through the origin, she merely said that they appeared to merge to become one vector without explaining how the situation modeled a set of linearly dependent vectors).

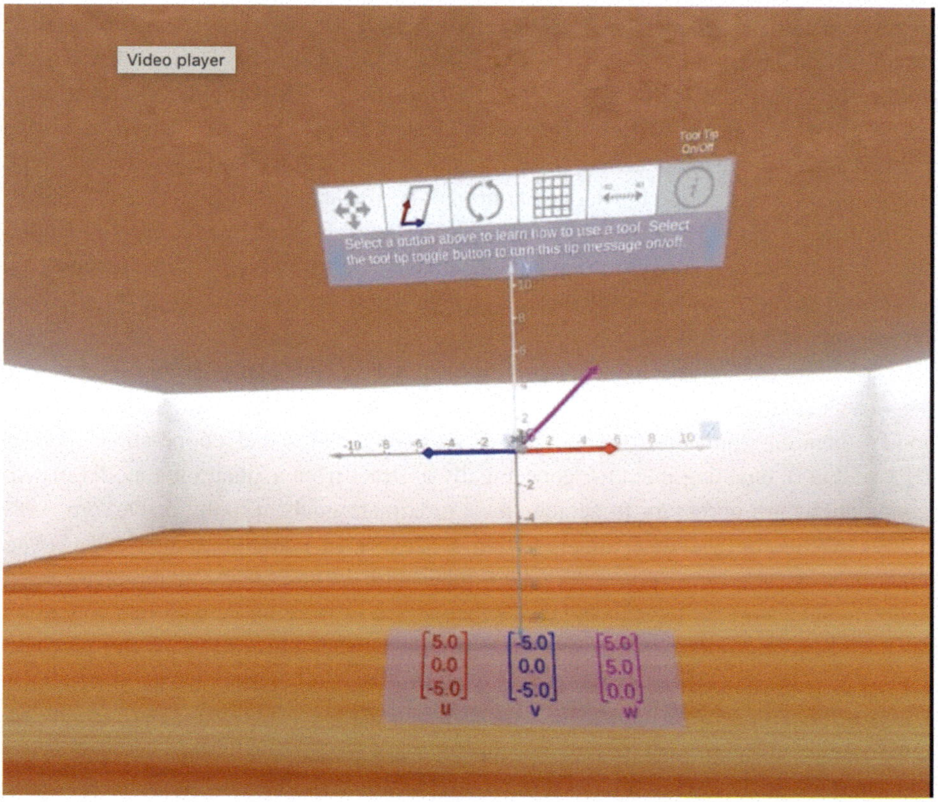

Fig. 1.11 VR vector manipulations module (Rivera et al., 2023)

1.3 Embedded Component in EC

In embedded cognition, the physical, social, and cultural environments could either constrain or support what and how we learn (Gallagher, 2023). Gallagher underscores the role of epistemic actions and cognitive niche constructions that emerge from our interactions with the environment where we think. To start with, learners as situated agents engage in goal-oriented *epistemic actions* that "actively manipulate the environment to reduce cognitive load (Kirsh & Maglio, 1994)" (p. 15). Croom et al. (2023) also associate *information-seeking actions* with epistemic actions, and these actions are different from *instrumental* or *pragmatic actions* that we perform daily to either achieve some outcome (e.g., walking to get to a particular location) or goal (e.g., communicating or pointing) or simply to "be creative and act for its own sake (e.g., dancing)" (p. 1). An instance of epistemic action in VR involves intentionally grabbing the line $y = x$ shown in Fig. 1.1 with the right hand and moving it up or down to help infer the concept that it is the y-intercept, k, in the equation $y = x + k$ that changes. In this particular situation within the VR environment, the epistemic goal of understanding the vertical translation of lines is linked with the epistemic action of moving the graph along the y axis, which fulfills the specific goal of acquiring knowledge about vertical translations of graphs. Also, the VR environment in Figs. 1.1 and 1.4 includes all the relevant mathematical representations (i.e., table, graph, and equation) that simultaneously change, which reduce the burden of interpreting them separately since they all change at the same time depending on how the learner grabs the line.

Learners as ecological engineers and problem solvers engage in *cognitive niche constructions* (Tooby & DeVore, 1987). That is, they interactively shape, interrogate, and modify their environment "through causal reasoning and social cooperation" (Pinker, 2010, p. 8993), enabling them to "conceptually abstract from a situation a model of what manipulations are necessary to achieve their proximate goals" (Tooby & DeVore, 1987, p. 209). Niche constructions "emerge from a network of continuous interplay between individuals and the environment, in which they more or less tacitly manipulate what is occurring outside at the level of the various structures of the environment in a way that is suited to them" (Magnani, 2007, p. 919). Furthermore, they differ from passive processes that result from natural selection and adaptation (Magnani, 2007), as the environment is seen to contain artifacts and other various tools that assist us in thinking, developing skills, and performing typical and intentional actions on them in order to complete particular tasks and also to meet our needs. Artifacts and tools function in the same way as internal mental representations (Hohol et al., 2021, p. 3). Consequently, we learn to adapt, solve problems, and create new solutions to problems (Gallagher, 2023; Magnani, 2007). Thus, constructing cognitive niches means using both artifacts and tools and "appropriate culturally transmitted practices" in ways that "make possible new forms of thought and reason" (ibid.). That is, acting on artifacts and tools may cause changes, modifications, or rearrangements in our behavior, including relationships and perceptions in the environment itself. Cognitive niche constructions also help explain the emergence of metaphorical abstraction, the "ability to readily connect abstract ideas with concrete scenarios" (Pinker, 2010, pp. 8996–97) and apply schemas acquired from one domain to another domain, "assembling them into increasingly complex mental structures" (ibid., p. 8998).

Some VR games, such as *Half-Life: Alyx* and *Powerwash Simulator VR,* allow individuals to engage in cognitive niche constructions. In *Half-Life: Alyx*, a resistance group led by a father-daughter team uses research to build tools that may be used to weaken and fight their invaders. While the game offers a limited number of weapons, players are able to upgrade and modify them in order to adapt and achieve timely improvements based on information they gather from their research. In *Powerwash Simulator VR*, players get to use different tools (e.g., power washer, different nozzles, extensions, soaps, etc.) with different cleaning purposes that allow them to rid a certain city of grime, algae, rust, etc. Depending on the task they confront, the cleaning activity engages them in adaptive gameplay experience. *Synthetic Selection* is an example of a non-VR game that exemplifies exploration, creation, and evolution. Players select from primitive pieces, including themselves, that allow them to evolve as a result of exploring some environment and the creatures that live in the environment. They engage in games that involve fighting with creatures that have adaptive behaviors. Figure 1.12 is a snapshot taken from the VR *True 3D* medical app developed by EchoPixel that allows surgeons to interact with VR images of CT scans, such as tissue or heart, to help them plan effective strategies during surgery. They are able to rotate the 3D pictures and generate cross sections to better see what they are dealing with (Erickson, 2017). Some VR-mediated medical environments, such as the *Neurosurgical Simulation Lab,* provide surgeons with manipulable 3D pictures of body parts from actual patients to practice on images for upcoming operations, enabling them to map out the situation with the patients, plan strategically and adaptively, and avoid potential disasters.

1.4 Extended Component in EC

Extended cognition assumes that our minds extend to our physical bodies and the artifacts or tools we use to accomplish tasks – that these external media have as much influence in our thinking as do our minds. Following Sprevak (2019), our minds are literally on those media; that is "human cognitive states and processes sometimes spill outside our heads and into objects in our environment" (p. 1). For instance, the XR apps that support the learning actions being modeled in Figs. 1.1, 1.2, 1.3, 1.4, 1.5, 1.6, 1.7, 1.8, 1.9, 1.10, 1.11 and 1.12 influence the way we think, how we behave and act, and the knowledge that emerges from this dependent relationship. Hence, our thinking has a physical basis that extends what we process internally in our brain.

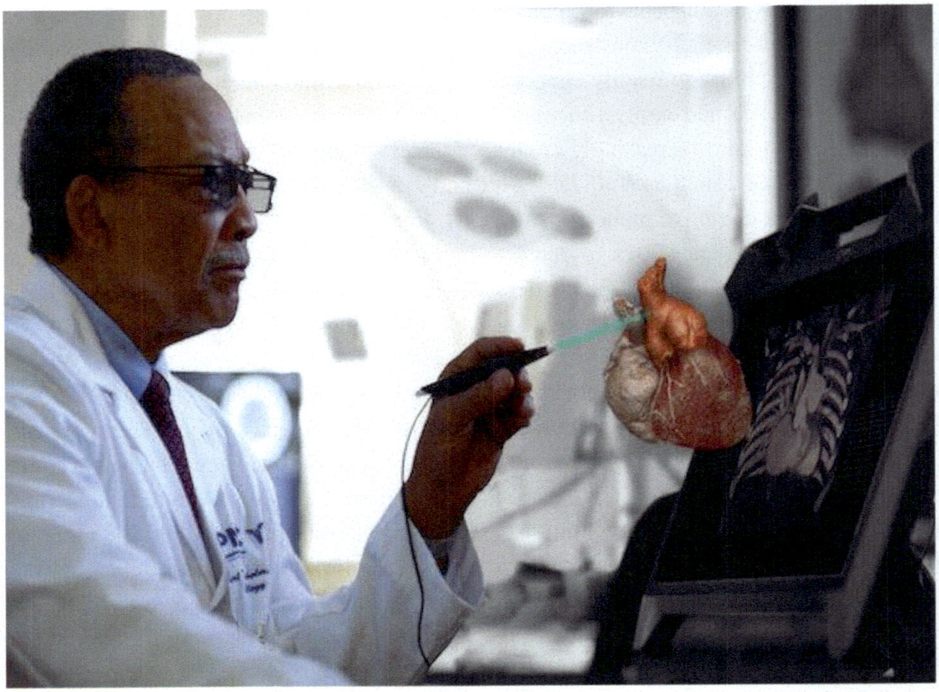

Fig. 1.12 VR true 3D (EchoPixel, 2021)

Gallagher (2023) underscores at least three waves or perspectives in understanding extended cognition. In this book, I only consider the first two perspectives since they matter to XR more directly than the third and other emergent perspectives. Across the different perspectives, however, extended cognition claims that when we think, our minds utilize mechanisms that go further than what our skulls and bodies are able to provide, which is consistent with the description provided in the previous paragraph. The first perspective draws on Clark and Chalmers' (1998) notion of *active externalism* in which our thinking draws on both environmental resources and our epistemic actions in activity. This particular processing demonstrates the *parity principle,* which states that physically manipulating an artifact (as an outer process) is "functionally equivalent" to activating the corresponding neural input or appropriate mental processing in the brain (as an inner process). The second perspective draws on Sutton's (2010) and Menary's (2007) notions of *complementarity* and *integration*, respectively. Complementarity means "different components of the overall (enduring or temporary) system can play quite different [functional] roles and have different properties while coupling in collective and complementary contributions to flexible thinking and acting" (Sutton [2010, p. 194] quoted in Gallagher, 2023, p. 22). Consequently, we model different degrees of complementarity depending on our individual internal processes and our respective environments, including access to the external resources. Integration depends on "reciprocal causal connections activated by the active bodily manipulations of the environment" (Gallagher, 2023, p. 23). Those bodily manipulations affect and are affected by physical and social environments, and include examples such as epistemic actions, biological/sensorimotor couplings, self-correcting actions (e.g., using language and tools to guide task completion), and cognitive practices (e.g., using normative and culturally-derived external representations such as diagrams and graphs to work through problems) that are integrated in the brain and mediated by our bodies. Consequently, engagement with external tools changes both our brains and practices (e.g., cultural, communication, social), which also means that what counts as internal and external changes.

What extended cognition means for XR is that what and how we think internally work in almost the same manner as what we do and how we think in the external environment, and this relationship is mutually determining and constitutive. That is, while aspects in the XR environment may cause us to think and perform actions in particular ways, we can also manipulate them in any way we want and still affect what and how we think. For instance, the virtually-drawn epistemic actions that surgeons develop and model, which they off-load to the *VR True 3D* shown in Fig. 1.12, help them practice the actions that would help them during the actual surgery. That is, the knowledge that emerges after a successful surgery is not solely attributed to the brain's internal thinking processes, but is a result of a tightly integrated brain-VR manipulation system. Certainly, XR apps on their own are useless; we need to interact with them (e.g., learn their functions) to help us think. Also, since the XR environment is incapable of exhausting all possible knowledge, we need to be careful about relegating all thinking to such spaces as if everything is a result of an environmental extension ("cognitive bloat").

1.5 Enactive Component in EC

Enactive cognition underscores the supporting role of thinking when we perform actions, which are grounded in sensorimotor skills (Engel et al., 2015). Actions span both overt behaviors and movements and inert processes such as thinking, calculating, and imagining. Actions also occur because there are goals to attain in activity. This action-oriented perspective on thinking shifts away from classical views that see cognition as primarily about "produc(ing) models of the world" that shape "thinking, planning, and problem solving" (ibid., p. 3) and toward the view of cognition as a "form of practice" (ibid., p. 1). Consequently, our cognitive states need to be understood in terms of how they relate to our actions. In particular, our brain states provide "directives that guide action embodied in dynamic activity patterns" (ibid., p. 5). Table 1.1 lists the classical and enactive perspectives on cognition that are worth noting and comparing in light of our interest in explaining how this enactivist view captures the importance of the actions we perform in XR activity. Thinking about the various XR apps shown in in Figs. 1.1, 1.2, 1.3, 1.4,

Table 1.1 Classical versus enactive views on cognition (Engel et al., 2015, pp. 2–3)

Classical view of cognition	Enactivist view of cognition
Our brains act as world-mirroring devices	Our brains act as vehicles of world-making that support, based on an individual's learning history, the construction of the experienced world and the guidance of action
Cognition involves computation over mental representations	Cognition involves the capacity of individuals to generate structure by action
The individual is a detached observer with a "bird's eye" view of the world	The individual is immersed in the task domain
Intentionality is explained by the representational nature of mental states	System states acquire their meaning through their role in action
Cognitive system structures are modular	Cognitive system structures are inseparable from embodiment
Processing in subsystems is context invariant	Cognitive systems need to be viewed holistically and dynamically, and processing is context-sensitive
Computations occur in a neural manner	
Cognitive models take into account only the inner states of individual cognitive systems	Cognitive models take into account the extended nature of cognitive systems

1.5, 1.6, 1.7, 1.8, 1.9, 1.10, 1.11 and 1.12 from an enactivist view, the generation and construction of knowledge does not reside solely "in the head" but reflects a dynamic interplay of brain, body, and environment systems (Gallagher, 2023).

In Chap. 2, we discuss the impact of XR apps in some aspects of learning, learning environments, and assessment in STEM. We also describe interesting results from research, and present a design framework that ensures the use of XR apps to support learning in STEM. Chapters 3 and 4 of the book explore how users' actions in virtual environments can productively support the development and construction of thinking in the STEM disciplines, including the implications of an action-oriented view of learning and other aspects of the 4E model of thinking in XR-mediated activity.

2 Extended Reality Effects on Learning and Assessment

In this chapter, I discuss the impact of extended reality (XR) apps on learning and other educational outcomes and how XR is reconceptualizing some aspects of learning, learning environments, and assessment in interesting ways. The findings related to the use of XR apps in educational settings have been consistently positive despite constraints with the design structures, concerns about their potential negative effects in our relationships with one another, and other perennial issues, which range from access for all learners and costs associated with scaling XR to unproductive behavior, physical discomforts, females experiencing cybersickness with VR headsets more often than males (Kelly et al., 2023), ethical dilemmas associated with safety and privacy (Georgieva et al., 2024), and a myriad of other concerns (Al-Ansi et al., 2023; Antonietti et al., 2000; Correa et al., 2023; Petrov & Atanasova, 2020; Shi et al., 2024). In Chap. 1, I explored how the 4E approach to cognition could be employed to support more meaningful acquisition of conceptual knowledge with the XR apps. The first part of this chapter provides evidence from several meta-analyses of XR apps' effects involving large samples of typically developing learners and learners that require additional support services. The second part of the chapter explores the potential of XR apps in supporting personalized learning and how to design assessments that are appropriate in XR-mediated learning activity contexts. The third and closing part presents a design framework for deep learning with XR apps. The framework offers one way to assist the field of XR in STEM education and research to develop common or shared elements when designing and implementing XR apps for learning purposes.

F. Rivera, *STEM Learning in Extended Reality*, Synthesis Lectures on Engineering, Science, and Technology, https://doi.org/10.1007/978-3-031-78973-1_2

2.1 Capturing Effect Sizes of XR Interventions

Several meta-analysis studies have established the effect size of XR on various aspects of education, in particular, learning. The XR involved in the studies (i.e., AR, VR, or MR), the outcomes being assessed, and the sample participants are detailed for each study. An effect size is quite interesting to know because it quantifies the strength of a perceived relationship between two variables, which typically involves an intervention and some outcome, or measures the magnitude of the effect of an intervention on an outcome in relation to a benchmark. Effect size values could be small, medium, or large, and the general idea is that larger effect sizes indicate stronger relationship between two variables. Of course it is important to be cautious about interpreting effect sizes in light of potential issues with empirical studies being utilized to determine effect size, such as choice of research design and methods, treatment duration, potential heterogeneous effects of moderating variables such as subject area disciplines. Also, in cases of descriptive or correlational investigations, an effect size should not necessarily be interpreted to convey that an intervention has caused the target outcome (REL West, 2021).

The overall analysis across different meta-analyses is that XR apps offer a powerful tool for improving educational outcomes across grade levels. Some of these studies are briefly discussed here to illustrate the consistent evidence of a medium to high effect size. The first is Chang et al.'s (2022) meta-analysis of 134 experimental studies (both quasi and rigorous) that investigated the effect of AR apps on three learning outcomes (response, knowledge and skill, and performance) between 2012 and 2021 (utilizing 201 comparisons between AR and non-AR instruction, 11,627 learners, elementary, secondary, and postsecondary foci, subject disciplines ranging from STEM to language, social studies, and other subjects, and treatment duration from 2 min to 1 semester with an average of 1.67 h). Response outcomes pertain to students' perceptions of the usability of the AR environment, attitude toward the learning activities, etc. Knowledge and skill outcomes align with the learning objectives that are being assessed. Performance outcomes assess their ability to apply knowledge and skills in real-world situations (e.g., conduct a scientific inquiry). The authors' analysis indicates that AR instruction had a significant impact on all outcomes with a larger effect size on performance, meaning to say that students in AR groups usually outperformed students in non-AR groups. In particular, the effect sizes for the response, knowledge and skill, and performance outcomes are $g = 0.49$ (nearly medium), $g = 0.65$ (more than medium), and $g = 0.74$ (nearly large), respectively. Also, AR apps that were used to learn language and social studies produced higher learner responses (e.g., on motivation or attitude) than AR apps used to learn science. An earlier meta-analysis by Garzón and Acevedo (2019) involving 64 quantitative studies ($N = 4705$) published between 2010 and 2018 regarding the impact of AR on students' learning gains (i.e., the improved learning outcomes between the start and end of an AR intervention) also indicates a medium effect size ($d = 0.68$). Interestingly, informal settings in which learning took place outside of the classroom produced a significantly larger effect size ($d = 0.80$) than formal settings ($d = 0.67$).

Tekedere and Göke's (2016) meta-analysis involving 15 pre-post studies (involving 1560 primary, secondary, and university students) that were conducted between 2005 and 2015 compared AR and non-AR groups in the following areas: success, attitude, motivation, learning level, satisfaction, efficiency, attention, interest, and trust. The general effect size across the studies was $g = 0.677$ (more than medium), which indicates that the AR intervention had a positive effect on the participating students. Kaplan et al.'s (2021) meta-analysis involving 25 pre-post studies (with participants over 18 years of age only) that were conducted prior to 2019 and covered all XR types (AR, VR, and MR) assessed whether XR-based training (e.g., simulation training environments) was as effective as traditional training methods on individual learners' ability to transfer the training, say, from the simulation to a target environment (i.e., assess the effect of XR training on some dimension of a real-world performance). The participants processed different types of tasks, from cognitive (i.e., using stored information that needed to be recalled or used later) to physical (i.e., assimilating learned psychomotor skills) and mixed (i.e., combination of cognitive and physical tasks). Their results indicate that both XR and traditional methods of training did not yield different outcomes and were equally effective at enhancing performance (with effect size of $d = 0.09$ on pre/post training). Interestingly, there were very small negative effect sizes in training effectiveness with the XR that would also indicate equivalent results for both groups. While the findings were not too exciting, the authors' following remarks about the value of XR apps are worth noting: "(T)he value of XR lies in providing training in circumstances, which exclude traditional methods, such as situations when danger or cost may make traditional training impossible" (Kaplan et al., 2021, p. 1). That is, "on the whole, [XR offers] a more valuable investment of time than traditional training methods" (ibid., p. 9).

Chen et al.'s (2022) meta-analysis involving 17 studies conducted between 2000 and 2022 (993 kindergarten to university students), assessed the impact of using XR apps on language learning outcomes. Their calculations produced a large effect size ($g = 0.825$), which was consistent with other similar comprehensive reviews in the area. It seems that the XR apps could provide a realistic simulated learning environment that could effectively support language learning. The authors also calculated the effect of XR in the following five language skills: speaking ($g = 2.1$), vocabulary ($g = 0.762$), grammar ($g = 0.590$), listening ($g = 0.495$), and reading ($g = 0.457$). The follow up qualitative research investigation also confirmed that the XR environment could provide a realistic environment for practicing speaking, enhancing vocabulary, and supporting vocabulary memorization skills. Grammar and reading skills would require more time and support from native speakers to produce noticeable improvements among learners.

Coban et al. (2022) generated Table 2.1, which provides a list of the effect sizes (see fourth column) of various types of VR on different educational outcomes drawn from different meta-analyses. The samples came from different groups, from elementary to high school and university students, adults, patients, doctors, specialists, etc. Overall, the authors note that on the basis of the results from Table 2.1, VR apps positively affected learning, skills development, and performance in significant ways. Interestingly, in a separate study that focused only on 48 primary experimental studies between 2016 and 2020 (3179 students, more higher education than K-12 participants) that investigated the impact of immersive VR apps on learning outcomes, their results indicate an overall less than medium effect on learning outcomes ($g = 0.38$). However, the subgroup analyses indicate a more medium effect ($g = 0.61$) on learning in K-12 and less than medium effect ($g = 0.31$) in higher education. Table 2.2 provides effect sizes for the different fields of education, testing type, measuring moment, control group type, and immersion type, based on the authors' more restrictive meta-analysis study.

The last set of meta-analyses involves learners that require additional support services. The overall impact of using XR apps to support and enhance their knowledge and skills is also very positive. For instance, Mosher et al.'s (2022) meta-analysis of 41 studies indicate better social skills (i.e., relationship skills, emotion recognition, social awareness, cooperation, executive functioning) for students with autism spectrum disorder (ASD) due to the XR intervention. Chen et al.'s (2022) meta-analysis of 112 studies that also employed different kinds of XR interventions improved ASD participants' "social interaction, acceptance, engagement, communication and speech, emotion recognition and control, daily living skill, problem behavior reduction, attention, cost reduction, anxiety symptom reduction, pretend play, contextual processing, match to sample skill, and insomnia control" (p. 104,682). In the area of learning, Denizli-Gulboy et al.'s (2023) meta-analysis involving eight single-case experimental design studies (21 participants, 9 in the 6–9 age group, 9 in the 10–13 age group, and 3 in the 18 and above age group; 15 males and 6 females) focused on the impact of AR apps on children with ASD. The overall effect size was very large (Tau U = 0.98), which indicates the potential of AR apps in effectively teaching new skills for this group of children. The authors also note that AR interventions for students with ASD "attract students' attention by visualizing abstract concepts and making the process fun" (p. 484). Furthermore, those interventions allow them to engage in their own independent learning at any time without requiring adult supervision.

Gu et al.'s (2023) scoping review of 66 empirical studies involving students with ASD identify strengths of and issues with XR for this group of learners. The authors provided more qualitative descriptions of the impact of XR apps' technical features and technology affordances on those students' autonomy (i.e., degree of freedom that the users have to think and act in an XR environment), human–computer interaction (HCI; i.e., how users dynamically exchange information in an XR environment via motion detecting, gaze tracing, and haptic feedback), and sense of presence (i.e., feeling of engagement in an XR environment, naturalness of the digital representations, attitudes toward the environment, level of immersion offered by the environment, and users' feelings of discomfort in the environment). The authors inferred that a little more than half of the studies provided

Table 2.1 Various effect sizes on the impact of VR apps on learning, skills, and performance based on different meta-analyses (Coban et al., 2022, p. 100452)

References	VR type	*k*	Variable (s)	ES	Outcomes/conclusions
Haque and Srinivasan (2006)	Simulator	16	Surgical performance	1.57 ↔ 2.18	VR Simulators reduce errors and increase performance in surgical tasks
Merchant et al. (2014)	Desktop	69	Learning	0.36 ↔ 0.77	Desktop VR applications are effective in learning
Guedes et al. (2019)	Simulator	14	Surgical performance	0.31 ↔ 2.67	VR simulator-supported training is more effective than Box-Trainer-supported training
Kyaw et al. (2019)	Simulator	31	Knowledge, skills	0.44 ↔ 1.12	VR technologies improve knowledge and skills compared to traditional teaching methods
Avcl et al. (2019)	Desktop, HMD	44	Learning	0.327	Three-dimensional virtual learning environments are more effective than face-to-face environments in terms of learning
Howard and Gutworth (2020)	Desktop, HMD	23	Social skill	0.69	VR is more effective than other educational approaches for improving social skills
Kaplan et al. (2021)	MR, HMD, Desktop, AR	25	Performance	0.13 ↔ 0.15	There is no significant difference between the education provided in virtual environments and other educational environments in terms of learning

(continued)

Table 2.1 (continued)

References	VR type	*k*	Variable (s)	ES	Outcomes/conclusions
Portelli et al. (2020)	Simulator, Desktop, HMD	24	Skill	$0.58 \leftrightarrow 1.34$	Virtual reality improves efficiency in the trainee's surgical practice and improves quality with reduced error rates and improved tissue handling
Zhao et al. (2020)	Desktop, HMD	15	Learning, satisfaction	$0.53 \leftrightarrow 0.77$	VR may be an efficient way to improve the learners' level of anatomy knowledge
Chen et al. (2020)	All types of VR	12	Knowledge, skill, performance time, etc	$0.01 \leftrightarrow 0.58$	Although VR effectively develops knowledge, it is not effective on other dependent variables
Jin et al. (2020)	Simulator, Desktop	28	Time, number of errors Scores	$0.99 \leftrightarrow 1.51$	VR is effective in laparoscopic surgery compared to current training methods
Woon et al. (2021)	Desktop, HMD	14	Knowledge acquisition	0.47	VR can be a viable teaching strategy to improve knowledge acquisition
Luo et al. (2021)	All types of VR	36	Learning performance	$-0.13 \leftrightarrow 3.10$	VR improves learning performance in K-12 and higher education
Wu et al. (2021)	HMD	35	Learning performance	0.24	HMD-based VR can improve both knowledge and skills development in terms of learning

k refers to the number of studies used in the meta-analysis; ES is effect size in *g*; HMD means head mounted displays

medium to high levels of physical autonomy that were related to the skills being targeted by the XR apps; high levels mean the students were permitted to actively explore, interact, and practice. Language autonomy was quite low due to the technical limitations with the XR apps; none of the apps did not provide machine-led conversations. Cognitive autonomy tended to be low to moderate, which means the students learned only what was

Table 2.2 Effect sizes for the subgroup analyses in Coban et al.'s (2022) meta-analysis (CAVE means Cave Automatic Virtual Environment)

Subgroup	k	g	ES level	Lower Cl	Upper Cl	Qb	P
Education level						5.96	0.015
K-12	79	0.61	Medium	0.38	0.83		
Higher education	26	0.31	Small	0.16	0.45		
Field of education[a]						52.98	0.00
Architecture	2	1.90	Very large	0.84	2.97		
Engineering	10	0.79	Medium	0.54	1.03		
Geometry	12	0.60	Medium	0.27	0.93		
Chemistry	4	0.55	Medium	− 0.21	1.30		
Language learning	10	0.49	Small	0.21	0.77		
Medical education	3	0.47	Small	0.09	0.85		
Geography	5	0.44	Small	0.28	0.60		
Safety training	18	0.34	Small	0.07	0.61		
Nursery education	2	0.24	Small	0.19	0.30		
Astronomy	2	0.23	Small	− 0.92	1.37		
Surgery	6	− 0.22	Small	− 0.42	− 0.02		
Biology	9	0.14	Very small	− 0.16	0.43		
Dental	2	− 0.14	Very small	− 0.46	0.18		
Anatomy	11	0.08	Very small	− 0.48	0.63		
Science	4	0.03	Very small	− 1.18	1.24		
Testing type						0.60	0.437
Knowledge	77	0.36	Small	0.20	0.51		
Task/practice	28	0.45	Small	0.27	0.62		
Measuring moment						0.95	0.33
Immediate	92	0.40	Small	0.26	0.54		
Delayed	13	0.27	Small	0.01	0.52		
Control group						3.90	0.048
Computer assisted	65	0.30	Small	0.16	0.43		
Traditional	40	0.54	Medium	0.30	0.78		
Immersion type						5.14	0.07
CAVE	5	0.41	Small	− 0.05	0.87		

(continued)

Table 2.2 (continued)

Subgroup	k	g	ES level	Lower Cl	Upper Cl	Qb	P
Education level						5.96	0.015
HMD	93	0.42	Small	0.29	0.55		
MR	7	– 0.17	Very small	– 0.75	0.40		

[a] Three subgroup variables (basketball, history, and computer) with fewer than two valid studies were eliminated from the analysis

part of their training and the script or pattern; they had little "control over their learning process" (Gu et al., 2023, p. 11). HCI approaches used mostly "keyboards, mouses, microphones, and touch screens" (ibid.) because they were easy to use, inexpensive, and pervasive. Also, most interventions employed computer screens (instead of head-mounted displays or smart glasses) for immersion in the virtual environment because they were "efficient, affordable, and convenient" (ibid.). However, some students became impatient and unmotivated, while others experienced motion sickness and difficulty interacting within the environment. Furthermore, the longer lengths of time in the environment caused more stress and prevented them from complying with the structures of the environment.

2.2 Supporting More Personalized Learning in XR Environments

Effective XR apps allow individual learners to experience personalized learning, which increases their motivation and engagement in their own learning. Personalized learning involves "customiz[ing] the learning experience for each student according to [their] unique skills, abilities, preferences, background, and experiences" (Herold, 2019, n.p.). Well designed XR apps that align with this view of learning are adaptive, flexible, accommodating of the needs and interests of individual learners, and can be adjusted based on ongoing user feedback. For instance, based on the 110 undergraduate humanities and engineering majors' responses on a survey regarding their experiences with VR apps, Antonietti et al. (2000) inferred that the students saw such apps as useful in supporting them when they had to "check immediately what [they] learned, to acquire a global vision, to facilitate understanding, or to apply the notions assimilated" (p. 326). The authors recommend the need to design "well-focused interventions" so that students "reject inadequate conceptions about VR or to realize educational properties of VR of which they are not aware" (ibid). Another instance comes from the world of work. Some XR apps are being used to train people to explore typical and challenging problem situations that can be safely explored and solved in a virtual environment. These apps help enhance training and learning in a synthetically-authentic and cost-effective manner. For example, Virtalis developed the VR app *ActiveMove*, which consists of immersive, hands-on, and experientially real 3D display systems with virtual instructors, to train the maintenance staff at the National Grid US in Boston, Massachusetts, USA to explore dangerous situations and learn ways to deal with them. Such virtual experiences are serious, engaging, and fun at the same time.

zSpace developed a tabletop VR that allows individual learners to explore a digital object 360^0 using 3D glasses that have a perception of depth functionality, a built-in tracking system, and a stylus. Using both the 3D glasses and stylus, they can drag objects in the virtual environment inside and outside the screen and investigate them in any orientation they prefer. zSpace (2023) also underscores several more ways in which XR apps can be used to personalize learning. In the list below, some of their statements have been modified to apply to all XR apps.

- *XR removes barriers between students and content.* Students interact directly with content by extracting virtual objects from a display, viewing them up close and in detail, and manipulating them using a stylus pen, [touch controllers, and hand trackers]. This inspires curiosity and allows them to "steer their own ship."
- *XR incorporates multi-sensory elements.* Using 3D glasses, [touch controllers, hand trackers], and a handheld stylus with haptic feedback adds visual, audial, tactile, and kinesthetic elements to immersive learning experiences, benefiting students who rely on one sense over others to stay engaged.
- *XR promotes creativity and critical thinking.* A safe environment encourages students to take risks, learn from mistakes, and think outside the box to find solutions. Research (Hite, 2021; Hite et al., 2021; zSpace, 2023) has shown that students who engage with lifelike [XR] content tend to ask higher level questions that help to deepen their understanding of abstract concepts.
- *XR builds confidence.* A setting where students are free to make mistakes boosts not only their confidence, but also their self-esteem and eagerness to go "above and beyond" in their learning, rather than stick to the bare minimum. Students who are risk-averse especially thrive in this type of environment enabled by [XR].
- *XR creates a collaborative classroom.* Because activities are problem-based, students are encouraged to work together in pairs or teams and find solutions. This collaboration helps the students build off one another's ideas, exposes them to alternate views and facts, and allows the social butterflies of the class to simply be themselves.

Petrov and Atanasova's (2020) study involving 80 high school students from Bulgaria used one of zSpace's applications to help them learn about the human cardiovascular system. Results of their pre-post VR-mediated experiment indicate significant gains in their students' understanding of the topic. They saw that learning with the VR app made teaching simpler. Furthemore, the students appeared very interested and motivated to learn the topic because they were given ample opportunity to investigate, experiment, practice, and "recover from failure" with the app (p. 9). They also note additional benefits such as "ease of use [and] high level of personalization," including teachers' capability to design instruction that "tailor learning for complex concepts" (ibid.).

While XR seems to offer increased personalized learning, we do not have sufficient empirical evidence to say that XR apps could offer a superior medium for either deepening content learning or supporting transfer of knowledge. For instance, Mayer et al.'s (2023) 20 years of immersive VR research did not produce "strong evidence that students learn academic material more deeply when it is presented in immersive virtual reality than when it is presented with conventional media" (p. 2236). They observed that some of those digital environments tend to be unfamiliar to students and "perceptually rich," which distract and prompt them to engage in "extraneous processing" (ibid.) that does not necessarily align with target instructional goals of a lesson. Ziker et al. (2021) note that for some students who used them, it seemed that there were gains in their content knowledge, however, the results could simply be attributed to a novelty effect (Clark, 1985). A third instance comes from Kang et al.'s (2020) experimental VR study with 312 undergraduate Singaporean students (187 in the control group and 125 in the treatment group) that indicates benefits in understanding and visualization skills for the treatment group. However, the experience did not result in better performance for them (i.e., they performed worse and not as well as the control group on some assessment items after using the VR app).

A fourth instance is drawn from Kaplan et al.'s (2021) meta-analysis of 25 empirical studies (samples of participants involved different groups of individuals [e.g., adults with autism, individuals with ailments, individuals from the general and typical populations, medical students, technicians]) that investigated the impact of using XR-mediated simulation-based training on transfer, that is, applying knowledge gained from training in XR on a task in the real world. The authors wanted to find out the extent to which the participants' learning in various studies that used XR "could fully transfer from practice to performance" (p. 707). Their overall results indicate that learning from XR is as "effective as traditional training approaches [such as those used in classrooms]" (p. 712) even if it seemed like XR apps offered "a more suitable medium for training on physical tasks [that target bodily training and psychomotor skills]" (p. 712) than cognitive tasks that involve remembering facts. They also note that the close similarities between the XR and the real environments supported the participants' ability to engage in transfer effectively. Furthemore, despite the equivalence result, they agree that the benefits of XR-mediated training (e.g., safety, costs, etc.) still make them a "valuable and superior option" in comparison with traditional resources and approaches, and recommend the need to attend to design and use appropriate tasks in ways that would encourage individuals to use XR apps even more effectively.

Ziker et al. (2021) has pointed out that usage of XR apps may focus "less on content knowledge and more on inquiry scientific thinking processes and behaviors and then collaborating and communicating" (Ziker et al., 2021, p. 63). For instance, Jiang et al.'s empirical study (2018) explores how a virtual environment that is loaded with several resources for supporting inquiry learning could affect students' self-regulatory learning (SRL) behaviors and strategies. The authors used a videogame-like immersive virtual assessment (IVA) environment to determine the extent to which 2,429 seventh and eighth grade students (ages 12 to 14 years) could develop science inquiry skills and SRL. Two inquiry situations were presented to the students, one involving frogs that developed six legs and another involving bees that were dying. The overall goal of the learning activity was to determine which among a finite number of possible factors might have caused each situation. In the IVA environment, they talked to virtual characters, read available information, made observations, and conducted various kinds of tests. They were also provided with a text-based-only digital notepad that recorded what they were thinking. The authors' view of SRL is metacognitive, that is, having "students develop an understanding of the learning task, set goals and construct plans to achieve their learning goals, execute various learning tactics and strategies, metacognitively monitor and reflect on their learning process, and adapt their plans, behaviors, and strategies accordingly" (Jiang et al., 2018, p. 29). The authors' first analysis focused on the students' action log data during immersion to assess whether those actions reflected SRL and whether the IVAs encouraged them to use SRL processes and strategies. Their second analysis tracked the students' note-taking and note-reviewing strategies in the virtual environment. Based on their data and analysis, they inferred that the students' experiences with IVAs "prepared them to adopt more efficient note-taking and note-reviewing strategies" (ibid., p. 28). They also "learned to exploit more available sources of information by taking and reviewing notes on them in order to either solve inquiry problems or monitor their solutions" (ibid). Furthermore, more time in the IVA context seemed to produce "more sophisticated" SRL behaviors and strategies, which could explain their "superior performance" such as "identifying correct final claims" that were supported with appropriate "causal explanations" (ibid., p. 47).

The preceding paragraphs might suggest an interest solely in individual learners and their experiences in an XR learning environment. One ongoing concern in XR STEM education research pertains to how such apps could be scaled effectively in classrooms and in ways that could support more collaborative learning among students. XR-mediated social learning research also appears to be in a developing state because the built-in capacity of XR apps to engage in co-presence and social presence, that is, "being there together and acting with other users at the same time" (Nguyen & Bednarz, 2020, p. 8), is still being developed. Some of the practical concerns about group processing in XR learning environments include the following:

- How to orchestrate meaningful and productive social inquiry and cooperative engagement among participants while using XR apps;
- How to create flexible XR apps that allow simultaneous multi-user play and co-experiences in virtual worlds (Battarbee & Koskinen, 2005); and

- How to foster and support effective group processes, collaboration patterns, and better relationships in the virtual environment over an extended period of time, including possible spillovers in their real lives and norms that are either preserved or emerge from the mediated reality experience (Nguyen & Bednarz, 2020).

Group size may also have an impact on "collaboration mechanism, communication, and social interaction between users, especially remote users, in XR environments" (Nguyen & Bednarz, 2020, p. 17). Based on their analysis of research into collaboration and group size, Nguyen and Bednarz (2020) note that large groups tend to affect communication and interaction and that a 3-person group size seems more desirable than a 6-person group size at least in the areas of "appropriateness, openness, and accuracy" (ibid.). Chiang et al. (2014), for instance, recommend using small groups to ensure that they "achieve greater efficacy in a manner that promotes the students' responsibility for their own learning as well as the learning of others" (p. 108). In the case of dyadic interactions, individual gaze and mutual gaze behaviors could support collaboration and conversation progress in situations when participants are negotiating for meaning.

2.3 Designing Appropriate Assessments in XR

Ziker et al. (2021) provide a good reflective summary of the recent assessments used in XR-mediated learning activity. At the very least, they remind us about the need to be mindful of potential dilemmas when we infer results from assessments generated in those settings. For instance, some students could experience discomfort when they use the equipment (e.g., head mounted devices in the case of VR apps). Other students might also feel hesitant about interacting with the apps and might prefer a traditional mode of instruction and learning. Furthermore, some students that are unfamiliar with them might need time to become familiar with the instructions, and still some others might get distracted by all the things they see in an XR environment. XR designers need to be mindful of accommodations and ways in which universal design for learning principles can be used effectively to deal with the needs of students. Ziker et al. (2021) also note that the preferred overall formative assessment mode seemed to favor more individualized feedback. Some studies also used traditional formative and summative items such as pre-post multiple choice questions to assess gains in learning, while other studies developed appropriate performance assessments that were inquiry-driven in structure.

How might we reconceptualize the nature and content of assessments in such environments? One possible approach is to use *stealth assessments* in a formative assessment context. Shute and colleagues (Rahimi & Shute, V., 2023; Shute, 2023; Shute & Rahimi, 2022; Smith et al., 2023) developed stealth assessments that involve collecting data in a

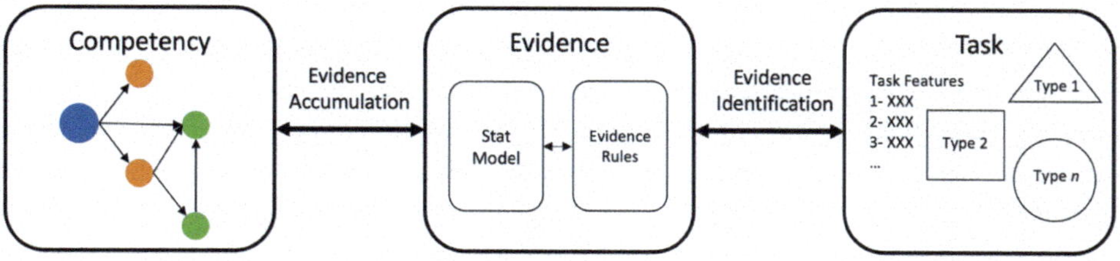

Fig. 2.1 Shute and Rahimi's (2022) evidence-centered design for stealth assessments

non-intrusive, automated manner and in real time while students are immersed in activity. Event logs, logged actions, and logged behaviors are identified, analyzed for patterns, and shared with students. The authors' stealth assessments use the evidence-centered design process shown in Fig. 2.1 with three interrelated components. The first component addresses a set of clear target competencies that will be assessed that are oftentimes unobservable in the beginning. The second component requires designing appropriate learning tasks that enable students to model the target competencies. The third component involves collecting observable evidence about the target competencies. Stealth assessments require collecting students' interactions that are generated when they use the digital tools (e.g., keystrokes, mouse clicks, and choice patterns) in order to monitor students' progress and provide ongoing feedback and support in real time. One issue with stealth assessments involves how to develop reasonably simple categories for coding large log files that are continuously collected (Ziker et al., 2021). Another issue involves "the generalizability of this approach and ways to further improve learning" (ibid., p. 4). Shute and colleagues, however, have also pointed out that various stealth assessments that were developed and used in various XR studies were able to validly assess learning in those environments (see Shute & Rahimi [2022] for a list of those studies).

To illustrate the evidence-centered design shown in Fig. 2.1, Chiang et al.'s (2014) experimental study is drawn upon, involving 57 fourth-grade students in Taiwan with a control group of 28 students and an AR-mediated experimental group of 28 students. The authors investigated the students' behaviors in a science inquiry lesson that emphasized making observations, inquiring, and sharing knowledge (as target competencies; first box in Fig. 2.1) in both AR and non-AR contexts. Inquiry fosters individual and collective exploration and discussion, which would enable the students to correct their mistakes and learn different perspectives. The location-based AR system contains the following five modules: camera (to take pictures), image editing (to edit or add text), discussion board, picture-based commenting, and discovery history (to allow access to what other students have shared). The AR system allowed the students access to all the pictures taken and shared by everyone in the group. Figure 2.2 visually summarizes the structure of the learning tasks (third box in Fig. 2.1). Students were asked to take pictures and to share their observations or ask questions using photographs, messages, or drawings. They received feedback from others as a result of sharing their pictures, and the goal was to encourage peer conversation among the students about their observations with no teacher interference, including their answers to questions and more questions that could lead to new explorations and learning. As evidence (second box in Fig. 2.1), the authors recorded the two groups' learning behaviors, movements in the real-life environment, and interactions with their peers, which they then analyzed using lag-sequential and quantitative content analyses. The authors found that the students in the AR group were fully engaged in the activity and interacted more with each other in constructing higher-level knowledge in comparison with the non-AR group.

Clarke-Midura and Dede (2010) developed *virtual assessments* to capture students' ability to engage in scientific inquiry, which involves processes of theorizing, questioning and hypothesizing, investigating, and analyzing and synthesizing. Their virtual assessments emerged from virtual simulations of real marine ecosystems, and students conducted experiments in different conditions to establish underlying causal models. The authors described a particular task that asked their participants to investigate the depletion of kelp forests. In the virtual environment that the authors designed, the students could take on an avatar character to move around the space, interact with the elements, and conduct experiments. In developing their virtual assessment, Clarke-Midura and Dede used

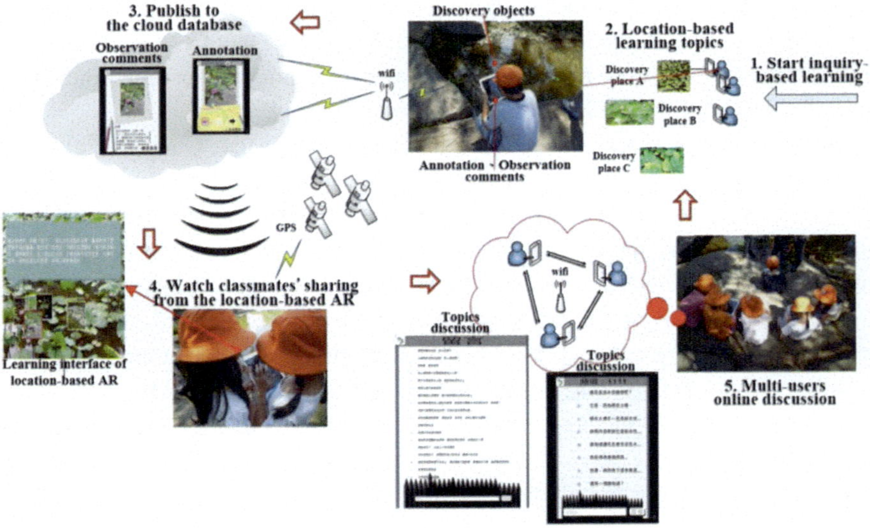

Fig. 2.2 Location-based AR learning task structure (Chiang et al., 2014)

a design framework that consists of the following five stages drawn from Mislevy and colleagues' evidence-centered design (Mislevy & Haertel, 2006).

- Domain analysis
- Domain modeling
- Conceptual assessment framework
- Implementation
- Assessment delivery architecture.

Domain analysis involves gathering pertinent information (i.e., content, concepts, terms, tools, representations, etc.) from teachers and other experts about a target domain, including the knowledge, skills, and abilities that will be the basis for establishing assessment claims and appropriate conceptions of the nature of knowledge that applies to the target domain. *Domain modeling* involves stating an assessment argument from the assessment designer that will link students' actions and various inferences about what they know or can do. *Conceptual assessment framework* (CAF) refers to the elements and processes that are needed to implement the assessment argument. The CAF generates an assessment blueprint that identifies: (i) an appropriate student model (i.e., knowledge and skills that are being assessed or measured); (ii) evidence model (i.e., behaviors that will elicit the knowledge and skills being assessed or measured); (iii) task model (i.e., situations or features of the environment that will elicit the evidence); and (iv) assembly model (i.e., the set of tasks that are needed to assess the breadth and diversity of the target domain). *Implementation* involves establishing the scoring rubrics, examples of items, item parameters, programming simulations, scoring algorithms, etc. The *assessment delivery architecture* involves the presentation parameters (e.g., whether students can use calculators) and scoring of items, including how they are stored in a task/evidence library (i.e., how to collect responses and their form).

One interesting aspect about Clarke-Midura and Dede's (2010) virtual assessments is concerned with how to evaluate skills that cannot not be simply assessed using standard assessments (like multiple choice items or hands-on performance assessments), which is important. The authors also see their virtual assessments as capable of eliminating typical concrete restrictions on assessments because "(e)verything [is] inside the virtual environment" (p. 324). Furthermore, while training into using virtual assessments is needed, scoring is automated, and the data and information generated can be quickly visualized in real time. Also, time and speed for investigating and collecting data in virtual assessments can be controlled, which can be fast forwarded or condensed if needed.

2.4 Designing XR Apps for Learning in Education

Based on recommendations from studies that employed XR apps for learning, including my insights that I have developed from my own projects, all XR apps need to explicitly address the following design components: (i) content/curriculum/task; (ii) app; and (iii) formative assessment. At the heart of this design framework for learning with/in XR is the perspective that such apps will be more effective, meaningful, and transformative for students if they are implemented and used in ways that align with the 4E—embodied, enactive, embedded, and extended—cognitive principles described in Chap. 1. Teachers who commit to using them primarily to demonstrate concepts for students, say, in a direct instruction learning environment, may not be fully maximizing the potential power of such apps in supporting more personalized and action-oriented learning in their classrooms (Kerawalla et al., 2006; Pang & Cai, 2023). Chapter 1 provides more details of the 4E cognitive processes that can be integrated in the design of XR-mediated learning activities, but the following three recommendations below are necessary.

- Develop XR apps that model strong embodiment, that is, by structuring knowledge and knowing through actions and interactions within the XR environment.
- Structure the XR apps in ways that enable students to model epistemic actions, which are intentional and aligned with domain-specific processes.
- Create situations in the XR learning environment that allow students to engage in cognitive niche constructions that draw on their epistemic actions in order to further deepen their content understanding and construct more metaphorical abstractions.

Aubrey et al. (2018) note that while students tend to be very enthusiastic about their learning when they use XR apps, they may not necessarily be learning significantly more with VR than with video or desktop games, or retaining more facts and transferring knowledge in comparison with those students who learn in non-immersive content. Hence, it is important to explicitly attend to ways that we can design effective XR apps that can support deep learning. Figure 2.3 provides a visual summary of a proposed design framework that can be used in designing impactful XR apps for learning in the STEM disciplines, whether the apps are implemented individually or in a social environment (i.e., in virtual worlds or classrooms). The arrows within each component and across the three components capture the complex, nonlinear, and nonhierarchical nature of learning in XR-mediated environments. Furthermore, meaningful cycles within each component and across the three components may produce deep learning, foster conceptual change, and support long-term memory.

The *content/curriculum/task design component* refers to the content and tasks of the XR apps, that is, the concepts and processes that will be modeled and the tasks that will be solved and analyzed. Domain-specific visualization is key, which is not the same as everyday ways of visualizing objects. Every STEM discipline has its specific way(s)

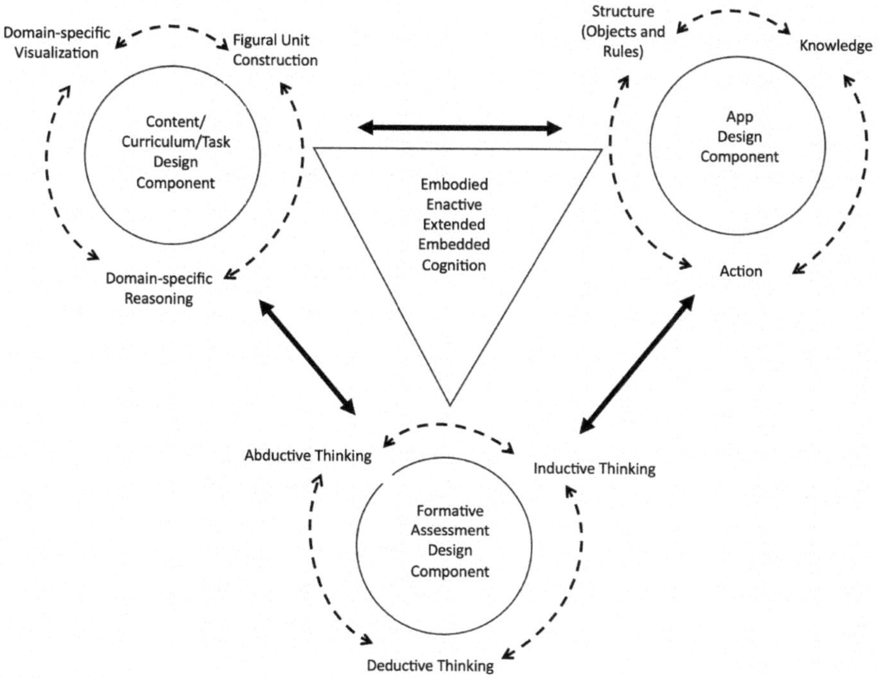

Fig. 2.3 Design framework for XR apps

of visualizing objects, but the initial process begins with being able to recognize the figural units or dimensions that comprise the structure of those objects. For instance, when the graph of $y = x$ is vertically translated 7.1 units upward along the y-axis from the origin in the *VR Transforming Graphs* app in Fig. 1.1, students need to be aware of its effect on both the y-intercept of the graph (as the figural unit) and the corresponding equation, which changes from $y = x$ to $y = x + 7.1$. Certainly, the goal is to model domain-specific visualizations that are accompanied by an explanation or reasoning. Using this design framework (across all STEM subject disciplines) the content/curriculum/task design component takes into account how learners are expected to acquire the target concepts and processes, and solve tasks as a result of initially interacting with an object's figural units in the XR environment. Otherwise, they will most likely feel confused and unable to maintain their attention on the learning or task at hand (Cabada et al., 2023).

The *app design component* refers to the trivium of structure (objects and rules for manipulating objects), knowledge, and actions that define the primary elements of the XR apps. The structure needs to be well-defined, simple enough to reduce cognitive load, and easily accessible through epistemic actions on the objects in order to generate emergent knowledge. All three aspects need to be coordinated effectively for productive engagement with the apps and in the virtual environment.

The *formative assessment design component* refers to the thinking moves that students need to model when they are using the XR apps. These are formative assessment moves that need to happen in XR-mediated learning to ensure that inquiry deepens into content knowledge construction, development, and proof. The trivium of abductive, inductive, and deductive thinking deepens the reasoning element in the content design and the knowledge and actions in the app design. The abductive thinking move encourages students to develop conjectures and generate tentative rules (e.g., "If I move the graph of $y = x$ three units up, then I see that it is recorded as $y = x + 3$" or "If I move the graph of $y = x$ five units down, the equation should be $y = x - 5$."). The inductive thinking move confirms an abduction (e.g., "Let me move the graph of $y = x$ up and down repeatedly to verify that the k in $y = x + k$ records the change that I see is happening with the graph."). The deductive thinking move tests an established rule on specific instances and situations (e.g., "Let me see the effect of k on other graphs of functions."). Abductive and inductive moves support the development of generalized thinking, while the deductive move supports growth in abstract thinking, which draws out the essential elements in a generalization, and cognitive niche constructions in situations of transfer. There is no order in which abduction and induction is expected to occur. Some adventurous users may start engaging in repeated actions as a way of establishing an abductive claim, which is then, of course, verified by induction. Meaningful deduction, however, emerges from abduction and induction.

Table 2.3 is an example of a checklist of items that will need to be considered in designing an effective XR learning app from both the 4E Cognition and Design Framework perspectives. Note that the checklist does not address design principles for building the structure of an XR app (e.g., requirements for constructing attractive design elements, user friendliness of the app, etc.) as readers can easily access general and XR-specific principles from published studies.

The following additional practical points below are also worth noting.

- XR-mediated learning activities need to align with stipulated content standards in order to be useful and intentional (Alalwan et al., 2020; Pang & Cai, 2023; Tzima et al., 2019).
- Interesting XR-mediated learning activities draw on problem situations that require students to use the XR app to analyze and solve them.
- Students may need to be trained on how to use the XR apps effectively in the beginning.
- There is no sufficient empirical research to confirm this, but students may benefit from more traditional or guided methods to learn a target content first and then use XR to deepen understanding. Students may also initially investigate a target concept or process on paper or use concrete manipulatives followed by a virtual engagement with an XR app.

Table 2.3 Checklist of items for designing effective XR apps for learning

Components	Does the XR app
Embodied cognition	• Model weak embodiment? (The body is not as important as brain; the brain is central to cognitive processing.) • Model strong embodiment? (Body-mediated experience; movement-driven; movement emerges from a soft assembly process; movement and XR environment co-evolve to produce meaning)
Embedded cognition	• Use mostly instrumental or pragmatic actions (everyday actions to support an outcome or goal)? • Use information-seeking and epistemic actions (intentional actions to acquire knowledge of a concept)? • Support cognitive niche constructions (creating new solutions to problems and engaging in metaphorical abstractions)?
Extended cognition	• Model the parity principle? (What and how we think internally work in almost the same manner as what we do and how we think in the XR environment.) • Contain aspects that shape how we think and perform actions in particular ways, including the ability to manipulate those aspects?
Enactive cognition	• Allow us to model actions that are grounded in sensorimotor skills? • Promote an action-oriented perspective on thinking? (Such thinking emerges from structures that have been derived from our actions in the XR environment. Knowledge about those concepts emerges from practice.) • Model the view that knowledge does not reside in the head but is a reflection of a dynamic interplay between our brain, body, and the XR environment?
Design framework	• Model domain-specific visualization? • Provide a coherent and unified experience that links structure, knowledge, and actions? • Support growth in abductive, inductive, and deductive thinking?
Knowledge retention	• Support short-term retention of knowledge? • Support long-term retention of knowledge?

- Design XR apps and learning activities in ways that avoid cognitive and visual overload. Unnecessary distractions in the interface (e.g., using too much color, busy background, too many digital elements) might slow down the knowledge construction process. In terms of time limit that will avoid simulator sickness, it is recommended that XR learning should not exceed 10 min for children and 20 min without a break for adults, or provide a 10–15 min break for every 30 min of use (Aubrey et al. 2018, p. 16).
- Build XR apps and learning activities that are interesting to students first. Engagement and interest seem to correlate positively with improvements in students' learning response, perceptions, and learning performance (Chiang & Liu, 2023; Di Natale et al., 2020; Cabada et al., 2023).

- Be mindful that deviations from a task in the XR environment may happen and when they do find ways to help students refocus their attention on the task (Chiang et al., 2014).
- Provide guidance on how students may summarize and organize their solutions and discussions (Chiang et al., 2014). Develop strategies for helping students "explore new solutions, continue in-depth analysis, and arrive at new questions" (ibid., p. 107). Perhaps include avatars, intelligent agents, or other technologies that could provide timely intervention and assist with knowledge sharing.
- It is always good to keep in mind Bailenson's (2018) two key guidelines for building good VR content, which could be extended to all XR, as follows: (i) See to it that the use of VR is essential (as an oversimplified response to his question, "Does the content need to be in VR?"); and (ii) Good VR should "feel great" (i.e., "Don't make people sick.").

3 Samples of Extended Reality Apps in Mathematics

In this chapter, a few XR apps that have been developed to support mathematics learning will be reviewed. In the discussion of each app, the elements listed in Table 2.1 are used to identify some of the strengths and constraints of each app with a note that the analysis is based on what could be inferred from the available descriptions. Based on the assessment of the related research literature in the case of mathematics, it is clear more small- and large-scale evidence of their effectiveness is needed, especially how the content/curriculum/task, app, and assessment components of the design framework in Fig. 2.5 can be aligned and developed in ways that support the simultaneous evolution of and growth in learners' thinking and reasoning and their intra-actions in the XR environment. Furthermore, more classroom-based implementation research studies around sustained use of such apps will also help establish their important, and perhaps necessary, role in supporting students' learning of mathematics.

The need for such apps may also be explained in terms of three additional benefits to learners that Bujak et al. (2013) describe in terms of the physical, cognitive, and contextual dimensions of AR learning, which can be extended to all XR learning. The physical benefit comes from the opportunity to actually manipulate objects (physical and digital), which engenders more natural interactions between objects and learners and between learners. The cognitive benefit allows learners to draw on their XR experiences to actively connect visual and spatial information and representations, and symbolic and abstract understanding in a progressive manner. The contextual benefit underscores ways in which XR environments can support "collaborative learning and facilitate personally meaningful experiences" (p. 536).

F. Rivera, *STEM Learning in Extended Reality*, Synthesis Lectures on Engineering, Science, and Technology, https://doi.org/10.1007/978-3-031-78973-1_3

Pahmi et al.'s (2023) systematic review of 23 published studies from 2018 to 2022 that focused on AR and mathematics education note additional benefits to learners such as overcoming "learning barriers, mathematical anxiety, and other cognitive problems" (p. 1). Many of the AR apps that have been developed in those studies were meant to support growth in students' spatial abilities and conceptual understanding, including efforts to increase their motivation to learn mathematics. Korkmaz and Morali's (2022) qualitative meta-synthesis of 20 research articles (grades 3–10 and university students), two master's theses, and two PhD theses published between 2010 and 2021 identify a few more properties of AR, some negative aspects of AR, and contributions of AR as they relate to mathematics, learning, and learning environment, which should apply to all XR apps. That is, AR apps were perceived as "interesting and attention getting, fun, beneficial/useful, effective, interactive, cost effective, practical/usable, portable, and [capable of] creat[ing] and enhanc[ing] visualization" (p. 14). AR apps helped "develop reasoning skills, spatial intelligence skills and spatial visualization, reduce cognitive effort and conceptual fallacies" and "make learning concepts easier" (ibid.). They also positively "contributed to [students'] sense of self-sufficiency" and "increased communication and active participation in the classroom," offered "multi-sensory learning environment for visually challenged students," "enriched the learning environment, "enable(d) 3D animation and a natural interaction with virtual objects," and "create[d] a sense of reality" (ibid.). Regarding the negative aspects of AR apps, they could "increas[e technological dependence," "caus[e] distraction and health issues for eyes," and might be "running out of power frequently" (ibid.).

3.1 Sample AR Math Apps

3.1.1 AR App on Fractions

Song et al. (2020) developed a single-player handheld game-based AR activity called *Ali's Math Adventure* to help primary school students in Malaysia (no age and grade level specified) practice their understanding of fractions in a variety of situations (multiple choice tasks, solving puzzles and problems). The app, developed using Unity 3D and Vuforia AR software development kit (SDK), is played on a handheld device with 2D touch screen capability. It is AR because the digital elements move within a fixed real environment. The Vuforia AR SDK allows the camera to track images and 3D objects in real time. The content of the app aligns with the stipulated grade-level math curriculum. To play the game on Android, the students need to answer questions about fractions to unlock and move on to new collections and levels of items. Figure 3.1 shows the interface of the AR game. The picture on the left side of the screen activates a fraction question and the right side lists the possible answers. When the students answer a question, they press the confirm button.

Fig. 3.1 Ali's math adventure handheld AR game (Song et al., 2020)

Ali's Math Adventure is an example of a game-based math AR app that is meant to make practice fun and engaging for primary school children. It is certainly more convenient to use in comparison with traditional paper-and-pencil practice worksheets since all the visual representations are embedded in the AR environment. Students are also able to work independently through a range of problems and receive instant real-time feedback on their answers. The implementation study that came with the development of the app primarily focused on users' experiences with the app. Each user was given about 15 min to test the app. Results of the post-survey were positive; the participants were generally satisfied with the app and its attractive structure, felt that the app was fun to use, and appreciated the game storyline and flow. Since the overall intent of the study was to gather user feedback on the design of the app, we could not infer with certainty if the various elements in Table 2.1 were met, and how. In terms of the design framework, the authors mentioned that the game difficulty "follow[ed] the syllabus listed in" the appropriate curriculum (p. 7), and the content focus on fractions seemed to emphasize part-of-a-whole understanding. It would be interesting to learn more about how the AR game and the tasks used in the game could support students' thinking and reasoning about fractions and operations, which would then allow us to assess how they modeled some or all the components noted under embodied, enactive, extended, and embedded cognition.

3.1.2 AR App for Exploring Polyhedrons

Another AR-mediated math lesson involves using markers on paper that generate three-dimensional virtual information for users. For instance, Auliya and Munasiah (2019) developed an AR geometry lesson for 15 Grade 9 Indonesian students that offered them an option to activate 3D objects and other virtual information when needed. Figure 3.2 shows two sample pages from an AR lesson involving polyhedrons that were developed using the Android app and Augment's open platform. The QR code marker on each sheet, when clicked through a camera, generates virtual information and a 3D rendering of the objects. According to the students who used them, they enjoyed learning about the related concepts, felt that the learning materials looked interesting, had fun with the interactive experience, and were motivated to learn more. No additional information about the interactive nature of the virtual information and how the students engaged with the lessons in either individual or collective contexts has been provided. Considering the elements in Table 2.1, it would be interesting to see how students engage with the tasks, the 3D figures, and with each other in ways that support deep understanding of geometric figures and their underlying structures.

3.1.3 AR App for Exploring Probability Concepts

Cai et al. (2020) worked with 68 junior high school students (ages 13–15 years) in China who learned about probability concepts in three AR-mediated lessons over six weeks. Thirty-three of the 68 students were assigned to the experimental group, which completed three Android device-dependent AR-mediated lessons either individually or in small groups of two to four students, while the remaining 35 comprised the control group. For the first two lessons, the students in the experimental group placed AR cards and papers in the camera's view, activating pictures of coins, dice, cards, and balls and generating the relevant sample spaces (see Fig. 3.3). The third AR lesson had students initially tossing a real coin in front of the camera that recorded the outcome. The AR app tracks the outcomes and records the probability of a toss in real time and in graphical form (see Fig. 3.4). For the authors, the third AR app was a time-saver because it conveniently recorded all the tosses and graphs the results at the same time, which allowed the students to infer a relationship between empirical and theoretical probability. The authors' results on the pre-post assessments indicate that the mobile AR-mediated lessons led to increased learning gains for students in the experimental group. Also, while both experimental and

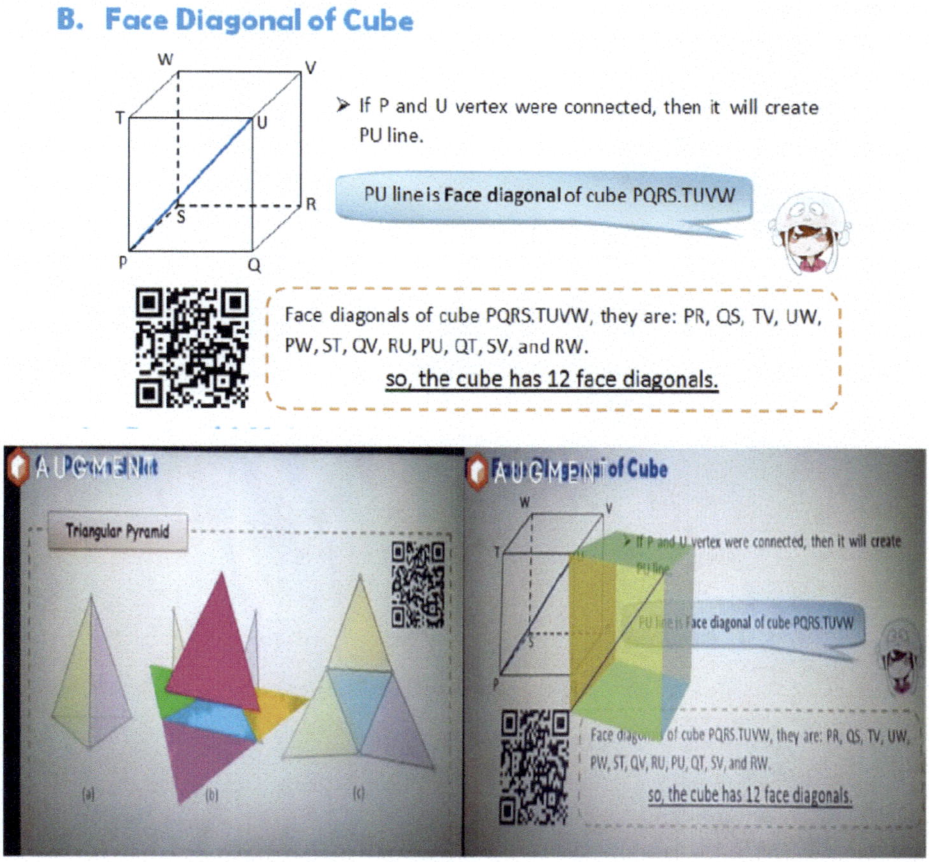

Fig. 3.2 AR for exploring polyhedrons (Auliya & Munasiah, 2019)

control groups scored higher on the posttest than on the pretest, the experimental group scored significantly higher on the posttest than the control group.

In terms of the elements listed in Table 2.1, the third AR lesson capitalizes on movement and students' actions (tossing, rolling a die), making it a good example of strong embodiment and enactive cognition. The students' actions were also intentional, epistemic, and repetitive that enabled them to infer a relationship between sample size and empirical and theoretical probability. Furthermore, when they performed the experiment in the AR environment, the virtual experience looked and felt the same way if they performed the experiment mentally. What was not reported in the study is the evolution of students' reasoning about each probability concept, especially how the discourse occurred between pairs of students and between students and their teacher. Overall, however, the lessons and the accompanying tasks offered a convenient, clean, and structured way of

Fig. 3.3 Activating a sample space in AR (Cai et al., 2020)

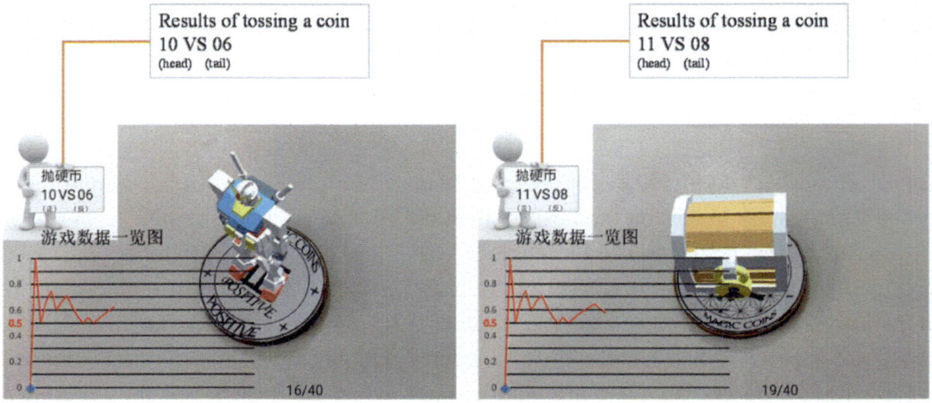

Fig. 3.4 Recording results of tossing a coin in AR (Cai et al., 2020)

performing all the three experiments without needing to use and track too many concrete manipulatives (sets of coins, dice, playing cards, etc.) to generate the needed outcomes.

3.1.4 AR App for Exploring Multivariable Calculus Concepts

Herrera et al. (2019) developed a mobile *AR in Calculus* (*ARC*) app to help students acquire spatial visualization skills in a multivariable calculus course. As of 2016, more than 900 students and instructors from different universities in Mexico have used the app. In one module, the students could author, manipulate, and visualize geometric models of surfaces in a 3D environment. Figure 3.5 provides a sample of two quadric surfaces based on the equations that were typed into the system. The students could rotate them from any angle and see them either as a solid or a mesh. They could also perform operations on them (e.g., obtain union and intersection of two surfaces) and see the results immediately. In another module, they were given access to ten sets of AR cards that activated different topics in multivariate calculus (e.g., quadric surfaces), and each card offered more opportunities to further explore a topic (e.g., manipulate 3D graphs by changing the coefficients of their corresponding equations). Overall, the students found the *ARC* app useful especially with respect to the topic on quadric surfaces. The authors also reported learning gains in their perceptions and spatial abstraction skills based on the pre-post tests. Some of the students' videos provided evidence of significant improvements in "their ability to describe surfaces in words [and] the type of graphics and drawings [they constructed]," which indicate improved spatial visualization skills. Furthermore, the authors observed that the students were able to grasp and apply related concepts quickly and clearly. In addition, the open and unlimited access to the app and interactions between teachers and students allowed them to construct correct explanations and focus on important aspects of the problems they were analyzing. Thinking about the elements in Table 2.1, the *ARC* app models strong embodiment and enactive principles. In the case of those lessons and tasks that involve 3D graph understanding, the students' ability to manipulate the graphs helped them develop the necessary mathematical structures through the epistemic actions they were modeling. It would be interesting to learn more about the quality and content of the students' reasoning and thinking while immersed in the activity; that aspect was not explicitly addressed in the study but was briefly inferred from the videos.

3.1.5 AR App for Exploring Hooke's Law and Galileo's Experiment

Jaber et al. (2024) developed two AR-mediated inquiry activities that provided a group of 3 Israeli Grade 11 students an opportunity to develop their conceptual understanding of covariational reasoning involving two quantities within the context of two real-life phenomena. Their study described in some detail how the students connected elements

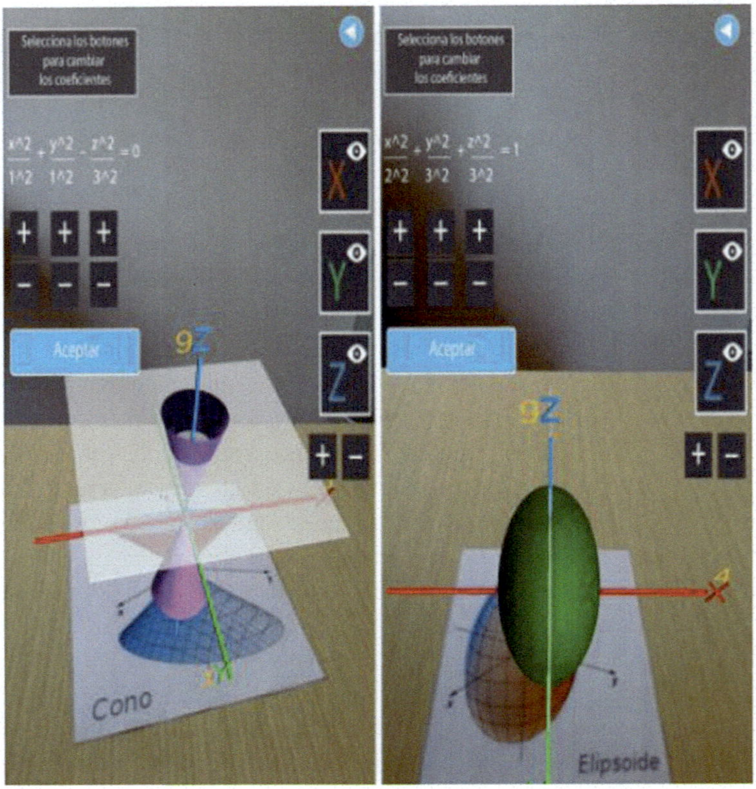

Fig. 3.5 Graphs of quadric surfaces in ARC (Herrera et al., 2019)

of the phenomena with their corresponding virtual representations. The experiment in Fig. 3.6 shows three students wearing an AR glass that allowed them to see the virtual representations (i.e., points and graph, table of values, and the x–y axes) and the materials used to perform the experiment (i.e., spring, stiffness of the spring, 50–100 g weights, and a bottle being filled with water). The experiment in Fig. 3.7 shows an actual cube sliding along an inclined plane with the same virtual representational structure (i.e., table of values, points and graph, and the x–y axes). The learning goal in each experiment, which lasted 3–4 h, was to establish the Hooke's Law formula, *length of spring* = *constant* × *mass*, in the case of Fig. 3.6, and the direct quadratic relationship *distance* = *constant* × $(time)^2$ based on the Galileo experiment in the case of Fig. 3.7.

The authors report that the students frequently connected real and virtual elements in their emerging understanding of (multiple levels of) covariational/functional thinking, and that their conceptual understanding could be attributed to "the potential of AR in juxtaposing [visually, in real time, both] worlds and bringing them to coexist" (n.p.). In terms of the elements in Table 2.1, the AR learning environment model strong

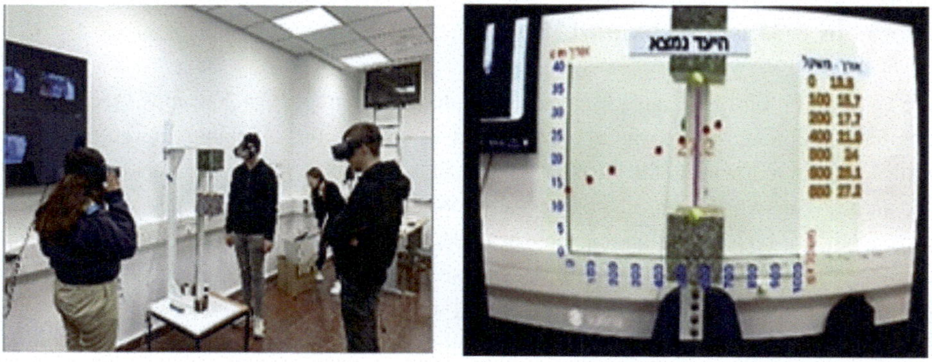

Fig. 3.6 AR-mediated experiment involving the Hooke's Law (Jaber et al., 2024)

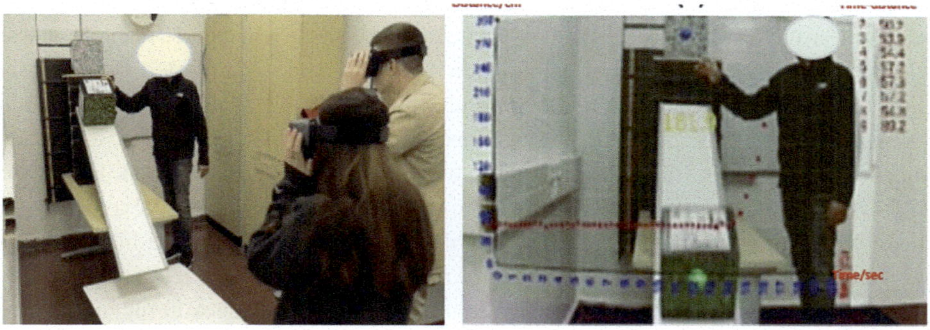

Fig. 3.7 AR-mediated activity involving Galileo's experiment (Jaber et al., 2024)

embodiment (since the students were manipulating and acting on real objects to generate data), strong embeddedness (since their actions were intentional and epistemic), extended (since there is a mutually determining relationship between real objects and visual representations and the students' actions on those objects and representations), and strongly enactive (since their conceptual understanding of covarying relationships emerged from their actions that guided their thinking). The design framework was well conceptualized, too, since the two experiments (i.e., the tasks) produced the target graphical representations of covarying relationships and their mathematical experience enabled them to link their knowledge and actions with the target structure of covarying relationships. The authors also described episodes that modeled how the students conjectured, experimented, and reflected throughout each experiment, which modeled aspects of abduction, induction, and deduction.

3.1.6 AR App for Exploring Different Mathematical Concepts in a Museum Exhibit

The last row in Table 2.1 addresses issues related to the impact of XR apps in supporting long-term memory. Based on a very limited review of the available literature on XR studies conducted within the last five years, it seems that more research is needed in this area. Also, while a good number of reported (pilot) studies usually state learning gains from pre-post assessments, they provide little to no information about how those gains were calculated, whether the tasks focused on higher order thinking skills, and how mathematical reasoning emerged in activity. Furthermore, it could very well be the case that the effect of those XR-mediated interventions on knowledge acquisition and retention is mostly short-term.

Sommerauer and Müller's (2014) experimental study on the effect of AR-mediated materials on short-term knowledge acquisition and retention in an informal learning context provides a good example of a rigorous and well-conceptualized study. Utilizing a pre-post test crossover field study, they wanted to determine whether 101 museum visitors (62 males, 39 females, age ranging from 14 to 79 years) were able to acquire and retain knowledge at a math exhibit with and without AR. They used Aurasma Studio to design twelve 60 to 252-s AR app exhibits that ran on a mobile AR app, tablet, and headphones, where 3 of the 12 exhibits were augmented with animations and the remaining 9 were augmented with video and audio of a curator describing and explaining the content (in comparison with the typical graphics and texts used in most exhibits). While the visitors were aware that they were part of an experiment, they engaged with the tasks without experiencing any intentional intervention. They also used the Aurasma mobile app on iPads to obtain access to the augmentations by using the camera to point at an exhibit, which would then trigger images. Their results indicate a medium effect for the study, and that the museum visitors significantly obtained more knowledge from AR-mediated exhibits than non-AR exhibits based on better performance on posttest questions related to the AR exhibits than posttest questions related to non-AR exhibits, and significant gains in their pre-post scores. Also, those who engaged in the AR exhibits thought that the AR was a valuable supplemental tool for presenting exhibit materials.

3.1.7 AR, 3D Printer, and 3D Modeling of Labyrinths and Mazes

Ulbrich et al.'s (2023) study is also worth noting. They used a combination of AR with Geogebra, 3D modeling, and 3D printing to design an activity that pre- and in-service teachers in Austria could use in their own classrooms, and it involves constructing labyrinths and mazes with certain mathematical features. The goal of the activity was to assist students in learning deeply about "computational thinking, combinatorics, reasoning, and results of geometrical operations" (p. 12), and become "motivate[d] to engage with mathematics" (p. 9). In developing the activity, the authors initially described cultural and historical information related to labyrinths and mazes, some mathematical principles (e.g., tilting, scaling, developing optimal paths) and design types associated with them, and their connections to the Austrian mathematics curriculum (i.e., fostering spatial thinking in second grade, perceptions of space and map drawing in third and fourth grade, and 2D and 3D formal geometry).

The authors implemented the activity with a group of 100 participants (preschoolers to senior adults) who were initially asked to investigate pictures of labyrinths and mazes found in architecture and other pieces of art to establish inferences about their geometrical and other mathematical attributes. From that firsthand experience, they constructed a 2D draft of their own design with additional conditions (e.g., their individual labyrinths should consist of squares and rectangles that only have one entry and a single object). Then they used their draft to construct a 3D version using Geogebra (see the series of pictures in Fig. 3.8). Next, they opened the 3D Geogebra app in their mobile devices (tablets, smart phones) and activated the Geogebra AR functionality, which allowed them to walk through the virtual maze and inspect and modify them as appropriate (Fig. 3.9). In the third and final part of the workshop, they downloaded their 3D Geogebra designs to generate an actual physical model using a 3D printer, which gave them another opportunity to mathematically assess their designs (e.g., make appropriate scaling to minimize 3D printing/production time) (Fig. 3.10).

Feedback from the 23 participants that responded to a questionnaire that was administered after the workshop indicates that they acquired new skills and learned Geogebra and the structure of 3D objects. They also appreciated the hands-on or artistic exercise and felt secure using AR and 3D printing. Referring to the elements listed in Table 2.1, this combination activity of 3D modeling, AR, and 3D printing is strong in all aspects of embodiment, embeddedness, extendedness, and enaction because of the strong design process that the authors engaged in prior to implementing the workshop activity. The activity and the tasks offer a natural or everyday context for engaging in purposeful reasoning, thinking, and acting mathematically since there is really "no technical knowledge [that] is needed to create [a] maze" (Ulbrich et al., 2023, p. 29). The authors shared a story of a 5-year-old participant who "was very joyful during the experience" (ibid., p. 27). He "grasped the concept of mazes and was still enthusiastic about it later since he started to create mazes out of building blocks for days after the workshop" (ibid., p. 28). His 3D maze is shown in Fig. 3.10 (in black background).

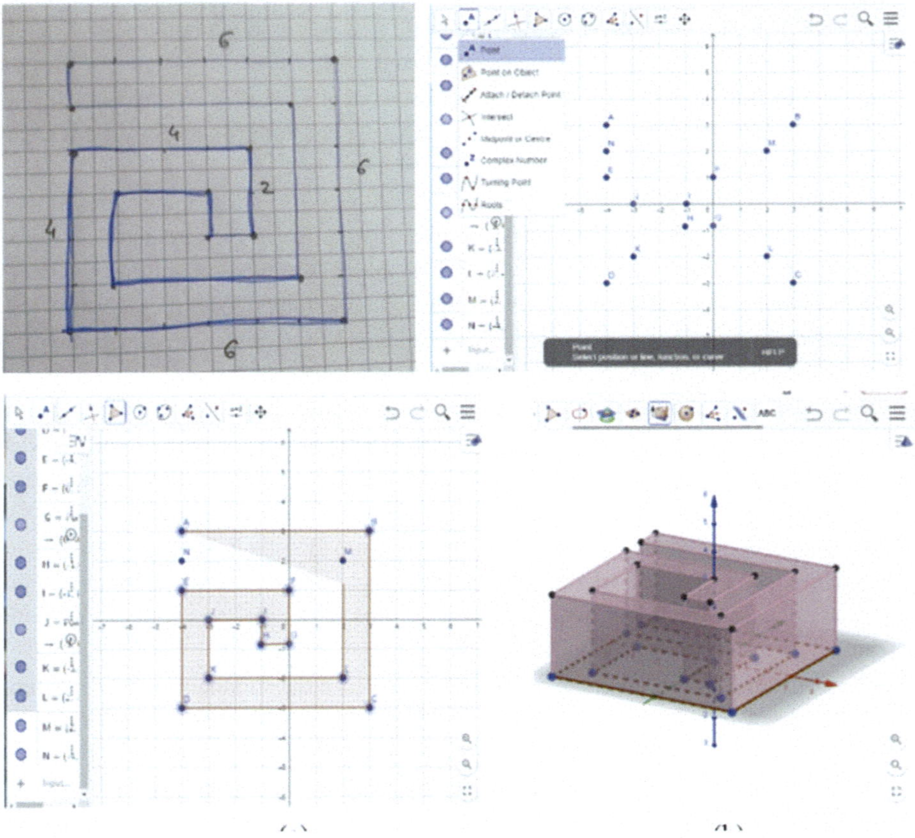

Fig. 3.8 Designing mazes from a piece of paper to 2D and 3D Geogebra (Ulbrich et al., 2023)

3.2 Sample VR Math Apps

3.2.1 VR App for Exploring 3D Surfaces in Multivariable Calculus

Herrera et al. (2019) developed a VR app called *AVRAM* (Remote Virtual Environments for the Learning of Mathematics) that helped more than 800 undergraduate students in Mexico visualize, manipulate (i.e., translate, change positions, rotate, scale, decompose, etc.), and investigate different aspects of a given surface (quadratic, planes) or surfaces in 3D space (see Fig. 3.11). The students interacted with their classmates and instructor, and generated regions of unions and intersections of surfaces that enabled them to correctly calculate their areas and volumes using integrals. The authors note that students' descriptions of those regions also transitioned from the informal phase using natural language to the more formal phase using inequalities. Results indicate that about 95% of

Fig. 3.9 Testing participant-generated mazes using Geogebra AR (Ulbrich et al., 2023)

the students found *AVRAM* useful in their learning of multiple integrals, quadratic surfaces, and tangent planes. Four years of student data also indicate improved final grades and lower failure rates in the course. Furthemore, their motivation and ability to draw and describe surfaces in words improved. Considering the elements listed in Table 2.1, the VR app seems to align with the characteristics noted under embodied, embedded, extended, and enactive (except cognitive niche constructions). What was not documented in the study was the nature of reasoning and thinking that the students modeled while they were analyzing the tasks with assistance from the app.

3.2.2 VR App for Exploring Basic Transformations of Graphs for Beginning Algebra Students

In this section, the VR app on transforming graphs of basic algebraic functions is described that was developed by myself and a graduate student using Unity. The *VR Transforming Graphs* app runs on any head-mounted device such as Meta 2.0. It has not been tested with students and in classrooms. Figure 3.12 shows the VR space that

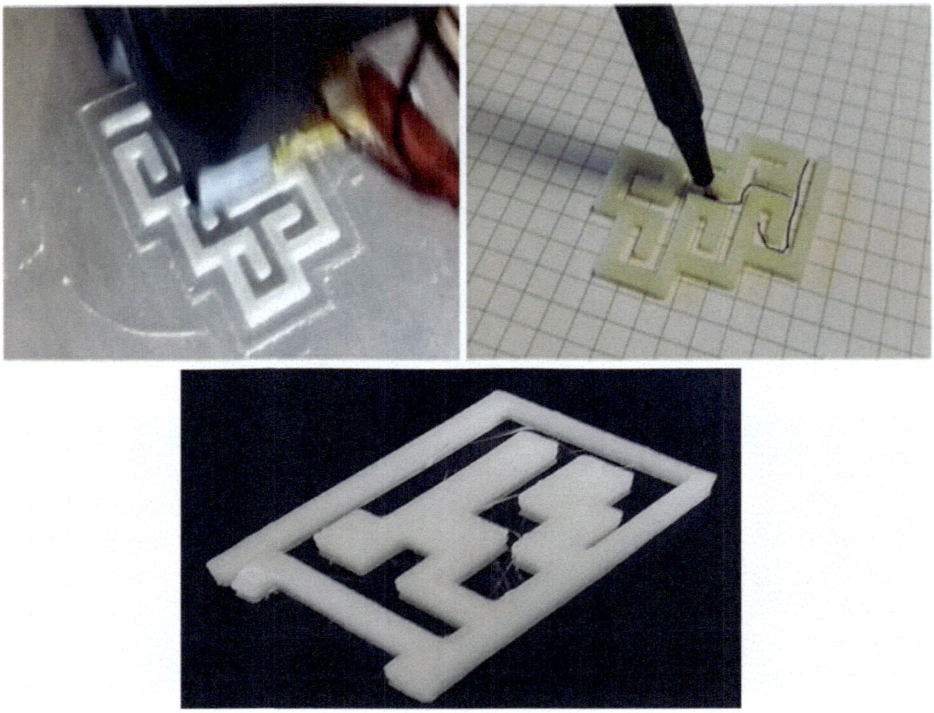

Fig. 3.10 3D printing participant-generated mazes (Ulbrich et al., 2023)

consists of a table of values on the left, the Cartesian coordinate system at the center, and the equation on the right. The history panel on the bottom left records the (series of) transformation action(s) performed on the parent graph with an additional button that when clicked would activate an animation of the (series of) transformation(s) from the first to the last step. The graph transformations panel on the bottom right consists of the equations of the parent functions at the top, the transformation actions, and buttons for resetting, for activating and deactivating the grid, and for moving near or far the board. In using the app, students have the option to use either the touch controllers or their hands to access the menus. From testing the app, while the touch controllers with their laser pointers are able to move graphs around, they do not provide the same embodied experience as using hand gestures to perform the mathematical transformations. Table 3.1 lists the graphical transformational actions and the hand gestures that students need to model in order for a graph to move anywhere on the Cartesian plane.

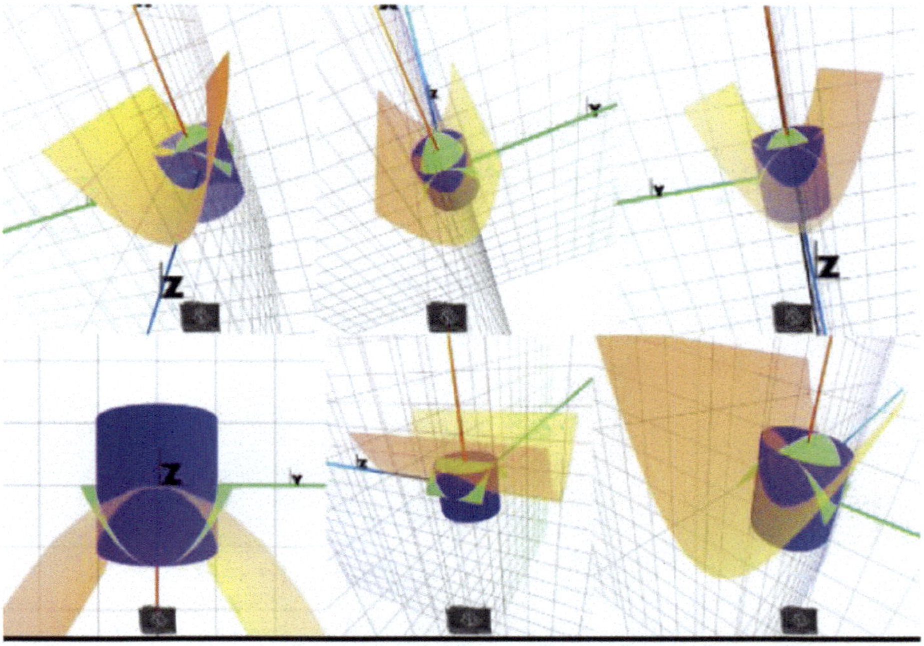

Fig. 3.11 Two students working in AVRAM to determine the volume of the intersection of three surfaces (Herrera et al., 2019, p. 19)

Using the parent absolute value function, y = |x|, theVR-mediated transformation is illustrated. Students need to first click the button for the parent function to activate its graph (Fig. 3.13a). Notice the three red dot points on the graph, which represent a sample of three important ordered pairs that are needed to obtain a sketch of its graph. Notice as well that the algebraic equation basically records the specific values of the parameters *a*, *h*, and *k*. To perform a horizontal translation of y = |x| several units to the right, the right (or left) hand forms a pinch and motions the hand to move the graph to the desired location on the right. A shadow of the graph of the parent equation stays in the original position, and the new graph emerges. This series of hand gestures would allow students to investigate and feel the effect of a horizontal translation action on both the graph and the equation (see Fig. 3.13b). The history notebook in Fig. 3.13c shows the recording of the different steps that were constructed to graph the equation

$$y = -2|x - 3| - 3.9.$$

The development of the *VR Transforming Graphs* app carefully attended to the elements listed in Table 2.1. When the functionalities desired in the app were planned, there was clarity about the need to develop a virtual learning environment that would model

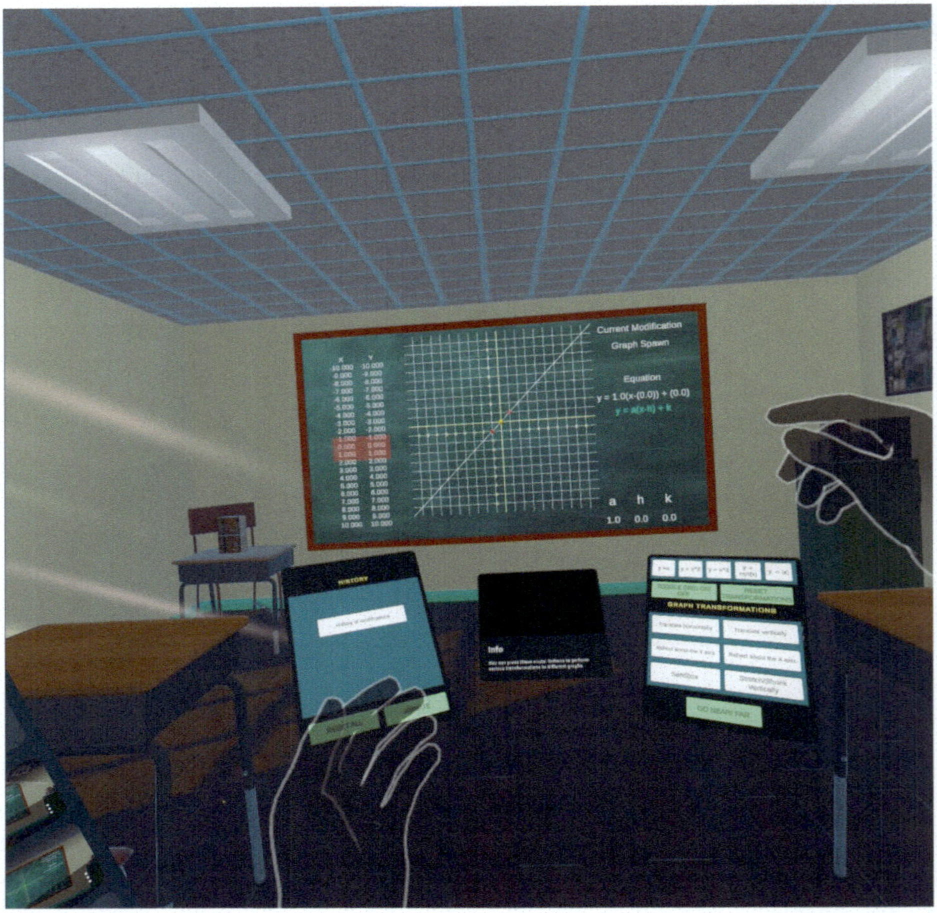

Fig. 3.12 VR transforming graphs app (Rivera & Paidi, 2024)

strong embodiment, embeddedness, extension, and enaction. For this app, all elements in Table 2.1 were attended to, with the exception of cognitive niche constructions, which is intended to be addressed more explicitly in a follow up VR app that focuses on problem solving.

In terms of the design framework, both the tasks and the content of the app align with the lesson on transformations of graphs for beginning algebra students. This lesson introduces students to the different ways of generating functions by performing various transformational actions on the parent functions. The permanent display of the parent graph and the graph of the resulting transformation(s) allows students to observe and determine what stays and what changes in both the graph and the corresponding equation. Also, many students struggle to graph functions correctly (e.g., in the case of $y = |x|$,

Table 3.1 Graphical transformational actions and their corresponding hand gestures (Rivera & Paidi, 2024)

When the transformation action involves	Do this
Reflecting a graph about the x-axis	Position the right (or left) palm horizontally along the x-axis and then make a 180° rotation
Reflecting a graph about the y-axis	Position the right (or left) palm vertically along the y-axis and then make a 180° rotation
Translating a graph horizontally	Move the thumb and index finger together to form a pinch and then move the pinched hand horizontally to the left or right
Translating a graph vertically	Move the thumb and index finger together to form a pinch and then move the pinched hand vertically up or down
Stretching or shrinking a graph vertically	Move the thumb and index of each hand together to form two pinches and • If the graph does not have a naximum or minimum point, move both pinched hands vertically in opposite directions to stretch the graph and toward each other to shrink the graph • If the graph has a maximum or minimum point, keep one pinched hand fixed and then move the other pinched hand vertically away from the extremum point to stretch the graph and toward the extremum point to shrink the graph
Rotating a graph clockwise or counterclockwise	Put the left (or right) finger on one part of the graph and use the pinched right (or left) hand to rotate the graph clockwise or counterclockwise

some students would calculate ordered pairs to the right of the origin, connect the points, and mistakenly conclude that the graph is a line). Hence, the three red points on both the parent and transformed graph indicate representative ordered pairs that would help draw the correct graph of the function. Additionally, the experience of seeing the graphs of the parent function and the transformed function helps students engage in figural unit construction and reasoning based on what they are able to infer from the transformation (e.g., the graph of $y = |x|$ includes a vertex, an imaginary vertical line of symmetry, and points on the graph that lie on the left and right of the symmetry line, etc.). Furthermore, when students model various epistemic actions on the graphs with their hands, the gestures reflect the intended transformational actions (e.g., motioning the two pinched hands to move away vertically stretches a graph). Finally, students have an unlimited opportunity to investigate the effects of each hand gesture repeatedly on any graph, which would allow them to engage in experimentation, observation, testing, and verification,

a

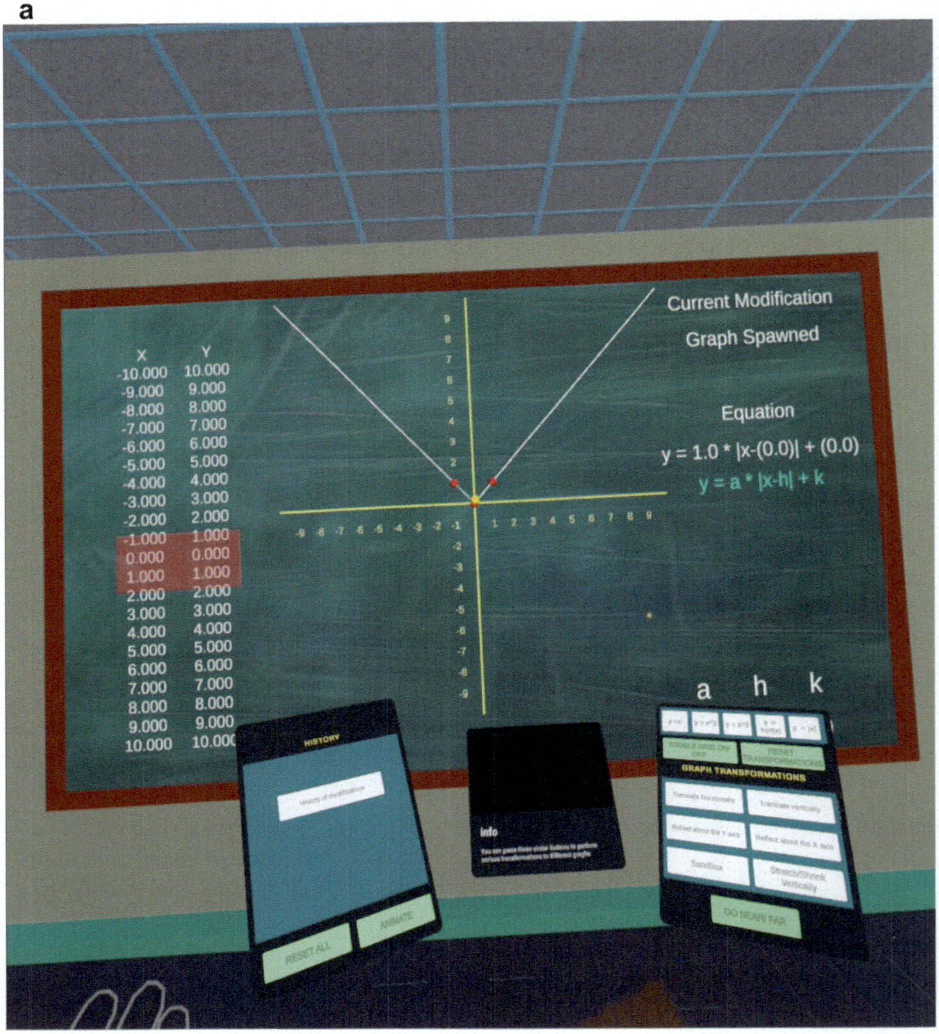

Fig. 3.13 **a** Graphing y = |x| on the VR transforming graph app (Rivera & Paidi, 2024). **b** Graphing a horizontal translation of y = |x| on the VR transforming graph app (Rivera & Paidi, 2024). **c** Graphing y = − 2|x − 3| − 3.9 on the VR transforming graph app (Rivera & Paidi, 2024)

and consequently produce abductive, inductive, and deductive statements (e.g., repeatedly motioning the pinched right [or left] hand to the left or right models the transformation action of horizontal translation and affects the variable *h* in the general equation).

b

Fig. 3.13 (continued)

c

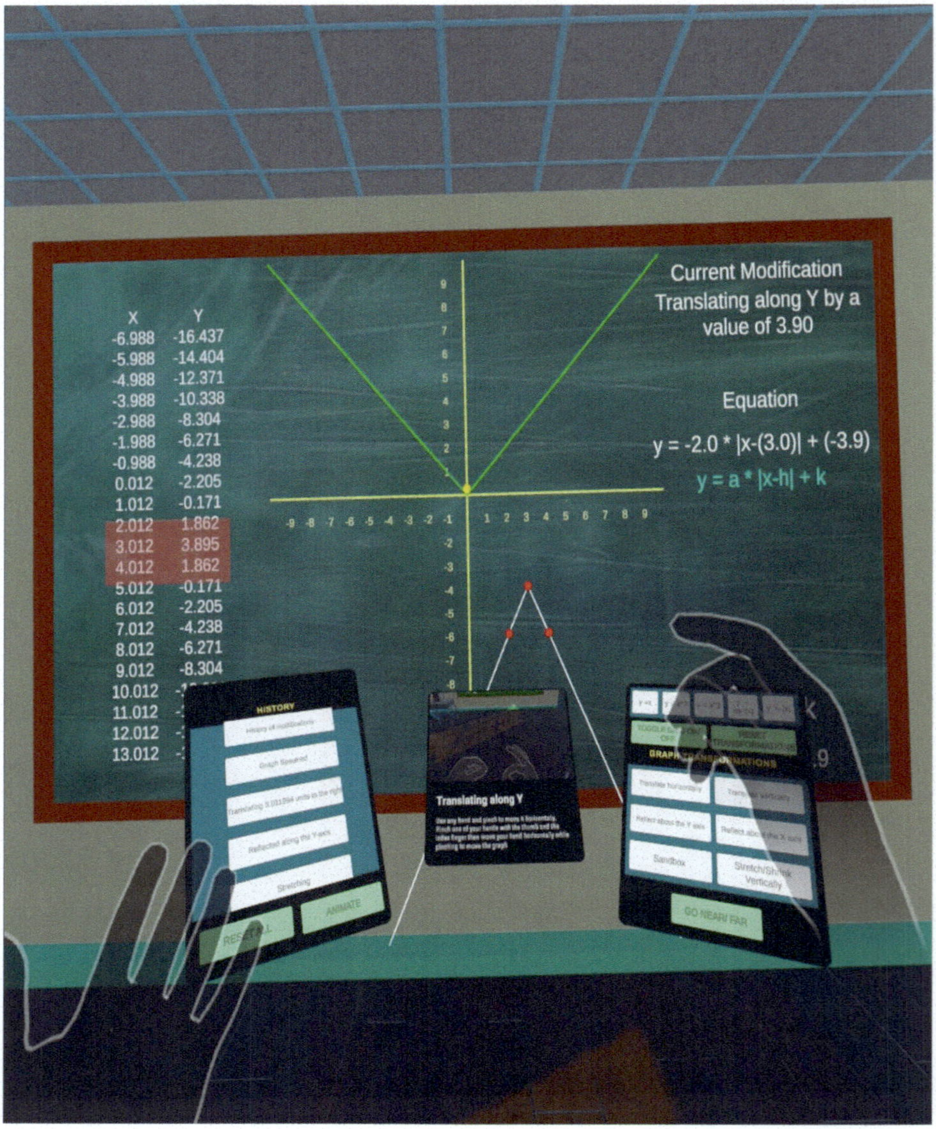

Fig. 3.13 (continued)

3.3 Mathland: An MR Math App

Khan et al.'s (2018) mixed reality (MR) app called *Mathland* supports "constructionist, simulation-based, and playful mathematical learning in the real world" (p. 1). Constructionist learning fundamentally involves building knowledge through engagement with everyday things. *Mathland* uses the Microsoft Hololens and two custom controllers that provide learners with a fully immersive experience. What makes *Mathland* different from VR, AR, and other MR apps, is the manner in which digital elements are constructed from their physical forms and the highly dynamic interplay of all elements in both the physical and virtual environments. In developing *Mathland*, the authors wanted to "merge the real and virtual worlds through real life-like interactions with virtual objects. We envision a world where people can interactively and seamlessly experience and play with mathematical phenomena in their real world" (p. 4). Users can, say, hold, move, and play with virtual objects. The physical environment also includes virtual annotations that facilitate "collaborative learning in the real world" (ibid., p. 2). The authors highlight the following three specific design components of *Mathland* below that allow users to virtually integrate physical elements in the real world into the MR environment (Khan et al., 2018, p. 3).

I. Spatial mapping for interactions with the physical world: the spatial mapping capability of the Hololens is used to scan the 3D physical environment and integrate it in the MR world. Integrating the user's physical world into the MR world allows virtual objects to interact with the physical environment like physical objects do, e.g. a virtual ball can bounce off a physical wall (see, for e.g., Fig. 3.14a).
II. Arm Controller for interactions with the human body: A custom wearable Arm Controller was developed to integrate the user's physical arm movements in MR. It allows the user to use their natural physical arm movements to interact with virtual objects in the same way they interact with physical objects. For example, users can play catch by catching and throwing the virtual ball in the same way they would play catch/throw with a physical ball (see, for e.g., Fig. 3.14b).
III. Object controller for physical manipulability: [An] Object Controller serves as a tangible placeholder or proxy for virtual objects in MR so users can reposition, rotate, and resize virtual objects in the same way as they reposition, rotate, and stretch physical objects. The user can gaze and tap on any of the virtual objects to connect the Object Controller to that virtual object. Any subsequent motion (i.e., a change in position and/or rotation) of the Object Controller is reflected in the position and rotation of the virtual object (see, for e.g., Fig. 3.14c).

The thirty subjects that participated in Khan et al.'s (2018) exploratory study found *Mathland* to be "engaging, efficacious in the areas of collaborative learning, problem solving, and mathematics education." Considering the elements in Table 2.1, *Mathland* seems to align with many of the elements listed under embodied, embedded, extended, and enactive

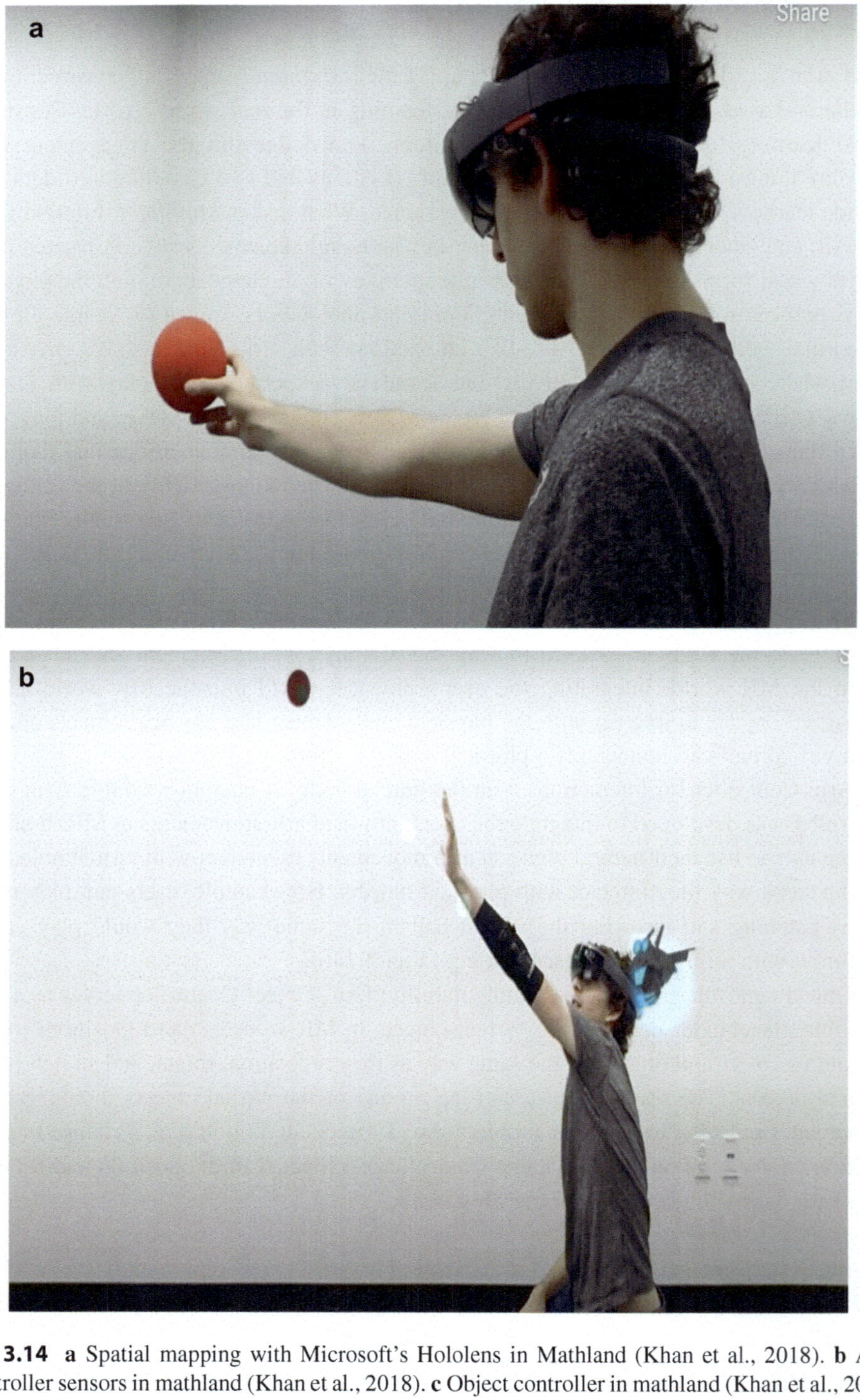

Fig. 3.14 **a** Spatial mapping with Microsoft's Hololens in Mathland (Khan et al., 2018). **b** Arm controller sensors in mathland (Khan et al., 2018). **c** Object controller in mathland (Khan et al., 2018)

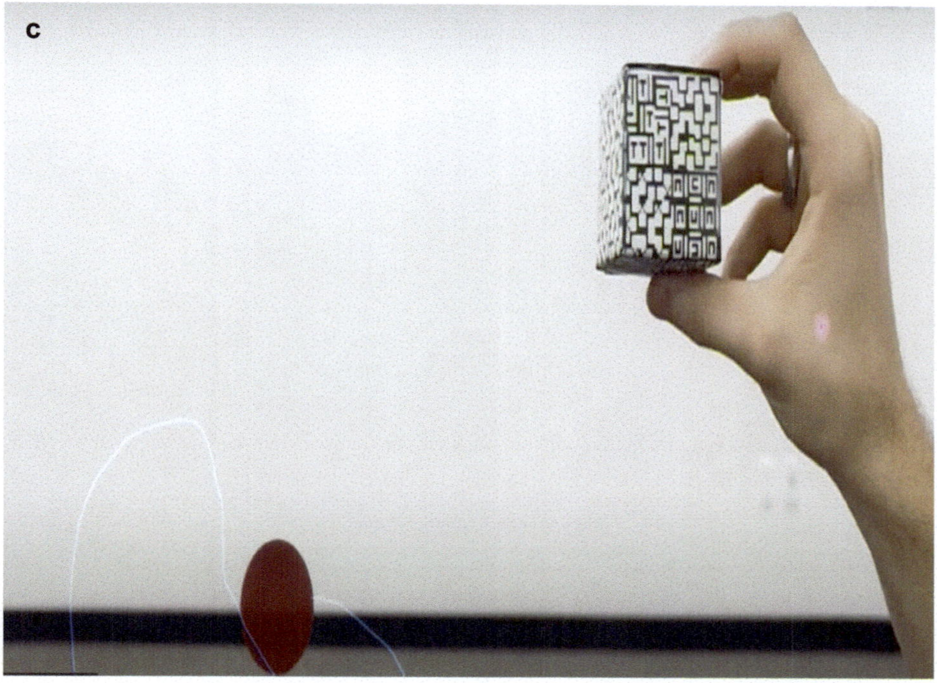

Fig. 3.14 (continued)

cognition. Based on limited information, it would be interesting to see how the MR app supports cognitive niche constructions and growth in mathematical reasoning and thinking from users' actions in such a highly immersive learning environment.

4 Samples of Extended Reality Apps in Science, Engineering, and Other Areas

The overall reception toward XR apps in the non-math STEM fields is generally positive and supportive. In particular, many descriptive and quasi and rigorous experimental studies and meta-syntheses in science and engineering that used XR as an intervention have consistently underscored the benefits of immersive technologies in reshaping education and training in these disciplines and, thus, strongly support more development and growth in light of those benefits. For instance, Yilmaz and Batdi's (2021) meta-analysis of 24 published studies (no specific details regarding total number of students, types of students, etc. were provided) between 2000 and 2017 that used AR to teach science "more concretely" and in a straightforward manner produced a medium effect ($g = 0.602$) on students' academic achievement. Tene et al.'s (2024) systematic review of 22 studies ($n > 1907$, no details about grade levels) between 2002 to 2023 that used VR and AR in STEM education also report their positive effects in enhancing students' performance, elevating their motivation to learn, and providing a more engaging learning environment (p. 1). Student performance "directly correlates with the essential competencies that [STEM] professionals need to develop and refine throughout their careers" (ibid., p. 5), and includes acquired and retained skills and knowledge, and proficiency and improved accuracy in applying those skills and knowledge. Student engagement refers to a "heightened readiness to incorporate these technologies into the learning environment, active participation in tech-driven educational tasks, and a persistent interest in the evolution of immersive tools" (ibid., p. 6).

Based on a reflective analysis of published XR studies in STEM (especially, science), there are some concerns that have been identified as important and necessary to address more explicitly in future work. Tene et al.'s (2024) recommendation below underscores the need to attend to ethical, multilingual/nontraditional, and inclusive immersive issues.

F. Rivera, *STEM Learning in Extended Reality*, Synthesis Lectures on Engineering, Science, and Technology, https://doi.org/10.1007/978-3-031-78973-1_4

> (T)he [VR and AR] field [as they apply in STEM] requires a globally inclusive and adaptable framework for rapid technological evolution and diverse educational contexts. Future research should broaden its scope to include multilingual literature and non-traditional academic channels, ensure ethical standards are upheld, and focus on personalization and adaptability to maximize the educational potential of these technologies. (Tene et al., 2024, pp. 1-2)

Matovu's (2023) qualitative analysis of 64 studies that have been published between 2016 and 2020 evaluated the design and implementation of VR apps in science learning contexts (biology, chemistry, physics, geology, environmental sciences, and general science; elementary to high school and university students; number of sample participants per study range from less than 50 students to more than 100 students). The studies employed VR apps to help students: visualize abstract concepts; enhance their learning experiences; help them practice their skills in conducting experiments; participate in virtual field trips; and engage in first-person experiential learning. While the authors acknowledge the students' overall positive ratings of their experiences in using the apps (i.e., engaged and motivated), they also identify several issues that can improve the development, and strengthen the implementation of, newer VR apps and, more generally, XR. At the very least, the design features, pedagogical models, and learning outcomes should be aligned and consistent to have a positive impact on students' content knowledge and experiences. For instance, direct teacher modeling with the VR without providing students with interesting storylines and sufficient opportunities to engage authentically in the environment, in either individual or collaborative contexts, will most likely limit their ability to visualize abstract concepts and model appropriate behavioral change and epistemic actions. Also, measures of content knowledge would need to go beyond mere recall of facts and learning experiences would need to be measured in ways that do not always depend on students' self-reports, which tend to produce positive ratings despite the fact that many of them might have experienced considerable difficulty using the app and learning in such settings.

In the following sections below, a few XR apps that have been developed to support learning in the other STEM disciplines are explored. Consistent with the analysis of the XR math apps in Chap. 3, the elements listed in Table 2.1 are used to identify some of the strengths and constraints of each app with a cautionary note that the analysis is based on what could be inferred from the available descriptions.

4.1 AR App on Wave Particle Duality in Physics

Cai et al. (2021) evaluated the impact of four inquiry-driven AR Optical Simulation Experiments called *AROSE* on an experimental group of 49 Grade 11 Chinese students' (ages 16–18 years) self-efficacy (i.e., degree of self-confidence to accomplish tasks) and conceptions of learning physics (i.e., "memorizing, preparing for testing, calculating and practicing tutorial questions, increasing knowledge, applying, understanding, and seeing in a new way" [ibid., p. 238]). A comparison group of 49 students used the three Flash-mediated experiments shown in Fig. 4.1, which consist of computer-generated screens with buttons that students can click to explore the demonstrations. The experimental group interacted with the four "intrinsically connected" AR-mediated experiments shown in Fig. 4.2 that the authors developed using Vuforia SDK/Unity3D. The authors note that "(f)rom the macroscopic level to the microscopic, from natural phenomenon to theoretical applications, AROSE can encourage students to acquire knowledge content, inspire discovery and exploration, apply knowledge, and promote a sequential and logical understanding of the material so that they have actively internalize them to a more meaningful whole" (ibid.). *AROSE* also provides an environment that offers real-time feedback and focuses primarily on the experiments and the inquiry process in order to help students attend to "high-level conceptions of learning physics than on low-level conceptions" (ibid.). Each *AROSE* includes a card that can be moved and rotated, contains areas that "can be pressed to control the lab equipment," has a slider that when moved can "change the size of the sliding rheostat" (ibid., p. 240), and has the necessary lab equipment with pre-programmed information that is activated when students perform certain operations on the card.

Results of the four-week intervention indicate that the *AROSE* group performed better than the Flash group in the areas of high-level concepts, conceptual understanding, higher-order cognitive skills, practical work, and social communication. *AROSE* also helped "reduc[e] students' cognitive load" and assisted them to learn "abstract content and concepts into lively and perceivable dynamic content" (ibid., p. 247). One participant said, *"By using this software, I can really see the electronic movement. Oh, I can't imagine it at all in the classroom"* (ibid.). Additionally, *AROSE* enabled them to "place themselves in experiments" that offered "a stronger sense of relevance and allowed them to interact with the virtual objects, enhancing students' sense of identity with the observed phenomena" (ibid.). Another participant said, *"In this AR software, I can actually change the parameters and see different movements"* (ibid.). Considering the elements in Table 2.1, while the AR seems to model weak embodiment and enaction, it is quite strong in both embedded and extended cognitive qualities. Furthermore, on the basis of the pre-post scores and students' feedback, it seems that there was synergy among the three design framework elements. That is, the content/curriculum/task and the app clearly emphasized what the students needed to visualize, and the inquiry process that was implemented to support the students engaged in reasoning, which, in fact, led to higher levels of conceptions of learning physics. It would have been interesting to see how the students modeled reasoning in all four AR experiments.

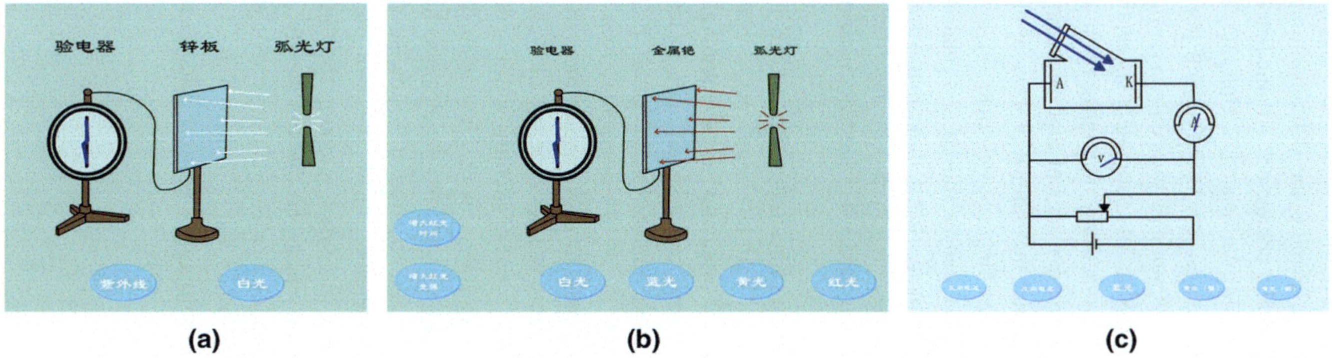

Fig. 4.1 **a** Experiment 1 simulates the macroscopic phenomenon of transmitting the photoelectric effect; **b** experiment 2 shows the microscopic phenomenon when the photoelectric effect occurs under the condition of power off; **c** experiments 3 and 4 present the microscopic phenomenon of the photoelectric effect under different current conditions (Cai et al., 2021, pp. 240–241)

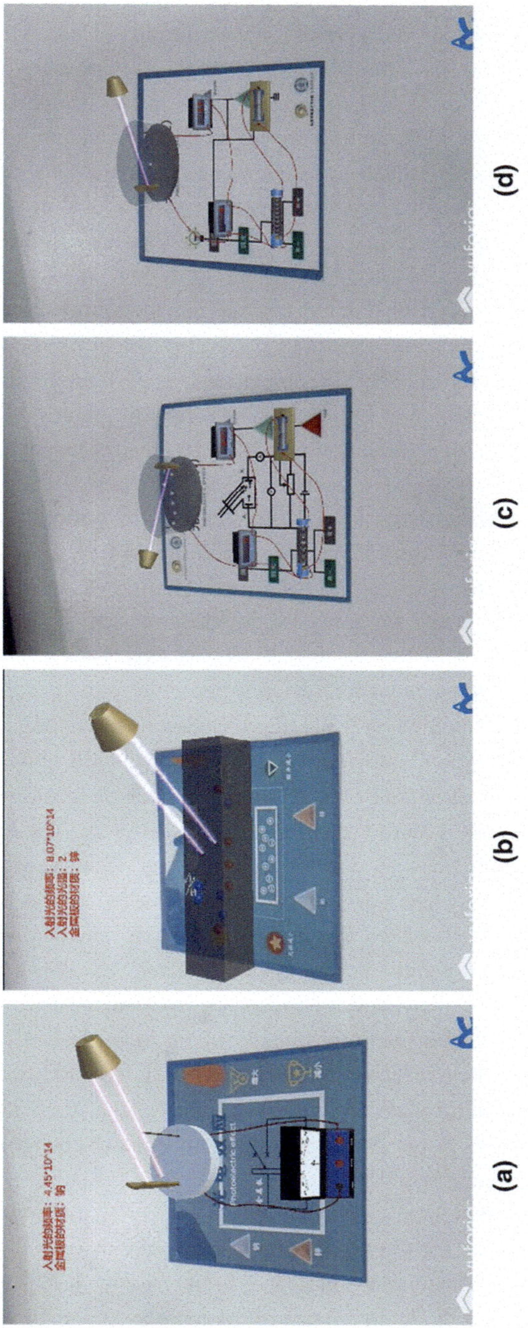

Fig. 4.2 **a–d** Four AROSE experiments (Cai et al., 2021)

4.2 VR and AR Apps for Understanding Climate Change

Newton et al.'s (2024) exploratory study involving two groups of undergraduate students (15 in the first group taking an Earth and Space science course and 8 in the second group taking a seminar on climate change and resiliency) investigated the impact of using VR and AR learning environments in how they conceptualized the scientific, political, economic, and sociocultural aspects related to climate change in the Outer Banks of North Carolina, USA. The authors documented the epistemic networks they created that supported them in their conceptualization process. The first group learned about climate change issues in the Outer Banks through a series of online videos and a follow up classroom discussion that focused on related science concepts and sociocultural issues. Next they used their smartphones and the *VRProTour* app to learn more about the target four locations in the Outer Banks. The VR app offered a 360° image of each location and contained necessary supplemental information via links to images, videos, and texts. Toward the end of the lesson, the students wrote and presented a management plan from a particular party perspective (e.g., as a business owner, scientist, politician, indigenous person, etc.). The second group experienced the same initial process of learning that was implemented with the first group. However, unlike the first group, the second group used the *Actionbound* AR app when they visited the actual sites, which activated links to important information that were triggered by geolocation. The students also engaged and interacted with relevant parties to discuss climate change issues. Then the students worked in pairs to write a management plan from a particular party perspective.

The authors' qualitative findings indicate high engagement among, and significant learning benefits to, both groups of students when they used the VR and AR apps in conceptualizing issues related to climate change relative to the target locations. However, the epistemic networks that the two groups created were quite different. The VR group "made moderately strong connections between technology and being engaged, while also making connections between technology and learning about the issue," while the AR group "made moderately strong connections between technology and being engaged, but made fewer connections between technology and learning about the issue" (p. 13). It seemed that the VR group benefited from "the increased feeling of presence in virtual environments ... of being in a place or environment while physically being located elsewhere" (p. 13). In the case of the AR group, they most likely benefited from "the additional exposure to the people and places impacted by climate change" (ibid.). Also, they were able to provide "contextualized examples of the physical impacts of climate change at specific sites visited as part of the experience" (ibid.).

Newton et al. (2024) did not provide a detailed description of how the VR group interacted with the elements in the digital environment, hence, there was no way of fully assessing the nature of embodiment that took place. In the case of the AR group, the combination of the actual visit and the strategic use of geolocator AR enabled the students to learn more deeply about the four locations, which made the experience strongly embodied for them. Furthermore, both the VR and AR apps supported growth in the students' epistemic actions, and the use of various resources (especially the videos and discussions) before and during the implementation phase provided useful information for modeling key aspects of extended and enactive cognition. In terms of the design framework, there was a clear synergy between content/curriculum/task, app design, and formative assessment that were all focused on the goal of the exploratory study, which was to develop the students' deep understanding of various sociocultural and scientific issues related to climate change. Hence, the type of thinking and reasoning that they modeled was consistent with the original intent in which the authors designed the VR and AR apps.

4.3 AR App for Solving Stereochemistry Problems

Habig's (2020) study investigated a group of 31 (16 females and 15 males) undergraduate chemistry majors in Germany who used AR representations to help them solve a 20-item multiple choice test (with single answers) involving concepts in stereochemistry. All the participants have already completed a basic course in organic chemistry, which was a requirement to participate in the study. Each test item included a ball-and-stick model of a chemical structure. Ten of the items used conventional 2D representations. The other 10 items used AR markers the author developed using Unity3D and Vuforia, which students could rotate and rescale using hand gesture controls (Fig. 4.3). All of them completed a survey questionnaire that asked them to rate their experience and attitude toward learning with AR, including a mental rotation ability test to determine whether gender (in this study, male and female) played a role in processing the test items.

The author's findings are interpreted with caution due to the small sample size. The test results indicate that while the scores on the 2D-based multiple-choice items showed no significant differences between the male and female students, the male students correctly solved more AR-based items than 2D-based items and the female students scored higher on the 2D-based items. The female students were not as successful as their male counterparts in solving the AR-based items, and Habig offered two possible explanations. First, he cites Castro-Alonso et al.'s (2019) meta-analysis involving 46 studies that shows larger effect sizes on the efficacy of dynamic (versus static) visualizations in studies that have significantly larger numbers of male participants. This indicates the need to explicitly attend to possible gender issues whenever designing XR apps. Second, based on the author's observation, the female students in his study might have found it difficult to simultaneously process textual and visual information while they were taking the test. This indicates the importance of providing all students with some initial training in using XR apps to help them practice and obtain strategies that help reduce cognitive load. Regarding the students' reception toward the AR representations, the author found no significant differences between the two gendered groups' responses. They saw the

Fig. 4.3 AR marker for a specific chemical structure (Habig, 2020)

AR models as providing a "meaningful supplement for 2D visualizations" and expressed a "high interest in learning with AR visualizations" since they "enjoyed working with them" (p. 639).

In terms of the elements listed in Table 2.1, while the students were able to rotate and rescale the chemical structures in the AR representations, there was not sufficient description in the paper to determine how the students' hand gesture controls modeled the components noted under the embodied, embedded, extended, and enactive cognition. With respect to the design framework, the AR representations seemed to align with the target content and tasks. However, the results and the structure of the assessment (multiple choice with single answers) did not provide sufficient description to obtain a full sense of how the students engaged in reasoning during testing.

4.4 Collaborative Immersive VR App in Geography

Jochecová et al. (2022) worked with 10 elementary to high school geography teachers (ages 27–55 years) from the Czech Republic who tested a collaborative immersive virtual environment (CIVE) module on hypsography. The goal of the authors' study was to

Fig. 4.4 Modeling in CIVE (Jochecová et al., 2022)

obtain the teachers' feedback on the lessons as experts, including their assessment of the eDIVE platform that was developed to support collaborative learning and teaching in a VR environment. The participants were divided into two groups of 5 experts in the beginning phase of the assessment. Then the whole group was divided into pairs much later during the collaboration part of the study. Figure 4.4 shows two pictures of participants wearing an Oculus Quest 2 head-mounted device and holding two controllers. Each room was also equipped with a computer that recorded the collaboration task in the CIVE. The assessment lasted two hours. Each group of 5 initially worked on geospatial tasks on eDIVE by following specific instructions.

In the eDIVE platform, multiple users are seen as avatars that can interact with each other in a virtual room. They may move around the room, explore content from different perspectives, and teleport as necessary. The room contains a table with a 2D map in the center (see Fig. 4.5a) that can be further modified and pictured in different ways (see Fig. 4.5b, c, and d). In designing the VR space for students, the authors wanted to provide both novice and expert students with an opportunity to develop their imagination, obtain easy visual access to complex phenomena involving different landscapes, and encourage them to move away from merely learning concepts by rote.

Jochecová et al. discussed one particular task called *Mirror Signals* that required the teachers to find a way to transmit signals using the fewest number of additional flags that could be placed on a map between two given starting and ending flags. Two kinds of instructions were provided to two different groups: one was thorough, structured, guided, and unambiguous, and the other was more general and open to encourage "more exploratory and creative ways of problem solving" (Jochecová et al., 2022, p. 4). Based

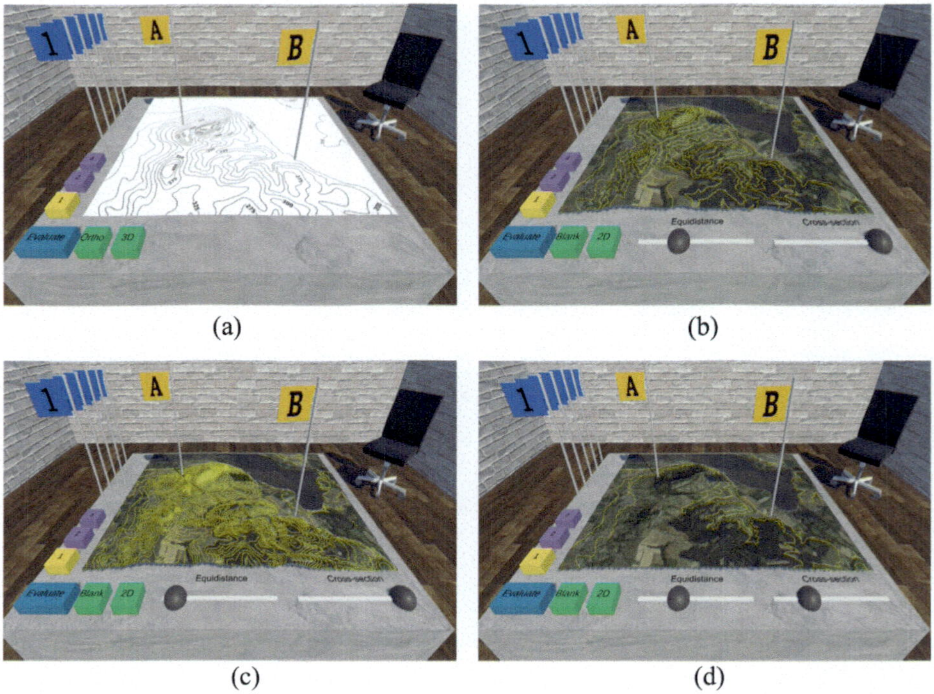

Fig. 4.5 **a** 2D blank map with contour lines and buttons that activate different functionalities; **b** 3D model with contour lines; **c** 3D model with changed contour lines interval by an equidistance slider; **d** cross-section of the highest parts of the terrain

on the teachers' assessment of the CIVE, students may benefit from more guidance. Furthermore, the CIVE may be motivating to students due to the interesting, "unbelievable," video game-like, and remarkable quality of the environment for exploring and problem solving. They also note that the CIVE:

- Has the potential to help students develop improved internal visualization and support understanding and deeper connections;
- Allows students to move and touch objects around on their own in order "to see what is happening" and "to understand how it works" (p. 8);
- Has animations that can replace the difficult work of drawing pictures by hand;
- Provides real-time immediate feedback and offers the feature of "going practically anywhere, both distant locations and other locations that are inaccessible due to the extreme weather—volcanoes, tectonic, or flooded karstic caves" (ibid.);
- Allows students to view actual places and "see the formation of riverbeds, karsts, or mountains and compare them according to their age" (ibid.);

- Allows students to participate anonymously, which will likely benefit shy users and provide equitable opportunities for all students;
- Supports students' developing ability to "create something and find out information themselves" (ibid.);
- Supports cooperation (with guidance); and
- May be used as a model for "explorative competitive collaboration" that may encourage students to work together through problems and seek solutions in the best ways possible.

Based on the authors' description of their CIVE app and the teachers' experiences when they worked through the *Mirror Signals* task, the app seems to address all the elements listed in Table 2.1. It is likely that the other geography CIVE modules target content that require students to model different types of strong embodiment, but the expected actions for carrying through the tasks in the VR space align with the embedded, extended, and enactive cognitive elements. It would be interesting to see how students' thinking and reasoning emerge in the CIVE context, including the extent to which they are able to engage in cognitive niche constructions collaboratively to support problem solving.

4.5 VR for Supporting Growth in Meta-Modeling Knowledge in Biology

Jansen et al. (2023) wanted to find out if prior instruction in meta-modeling in biology has an impact in how students apply meta-modeling knowledge in contexts that involve both 2D and 3D models. Meta-modeling knowledge in biology refers to levels of students' reasoning when they deal with the nature and purpose of models and multiple models, and engage in testing and changing models. The levels range from initial and basic (model object) to medium (model of something) and to expert (model for something).

Working with a class of 88 Dutch Grade 11 students (ages 16–18 years old), 41 students received training in a previous academic year to construct 2D biological models and engage in model-based reasoning that focused on the nature and purposes of simple and multiple models, while the remaining 47 students did not receive the same training and background content knowledge. Three lessons on the blood-glucose regulation process were implemented. During the first and third lesson, they responded to pre- and post-test items, respectively, that assessed whether working with a VR model influenced the students' expressed meta-modeling knowledge. During the second lesson, they learned about the regulation process in a VR-mediated context using their mobile phones that were inserted into a Google cardboard. In Fig. 4.6, the VR activity initially starts with a person in the virtual space eating a donut. The students then see the glucose molecules traveling "from the small intestines, through the cell membrane of the small intestines, into the bloodstream and into a somatic cell" (ibid., p. 66). While in the virtual environment, they respond to multiple choice questions that ask them to either explain what they see or predict the next step in the process. Also, each time they provide an answer to a question, they receive feedback with an explanation. The feedback process does not stop until they have chosen the correct answer.

After the students have completed the VR activity, they responded to the following homework tasks that included three 2D models and 1 VR model of blood-glucose regulation in varying levels of abstraction: name and categorize the differences between the four models; match and explain the purpose with the correct model; explain if either a 2D or VR model alone could meet all the purposes; name the choices made by the model creator to explain the purpose of each model; and whether one model could sufficiently describe the blood-glucose regulation process.

The pre- and post-test results indicate that students with the previous year background on modeling produced higher levels of expressed meta-modeling knowledge in comparison with the VR group only with no such previous background. The authors suggest that "teaching students about aspects of meta-modeling knowledge and letting them work with

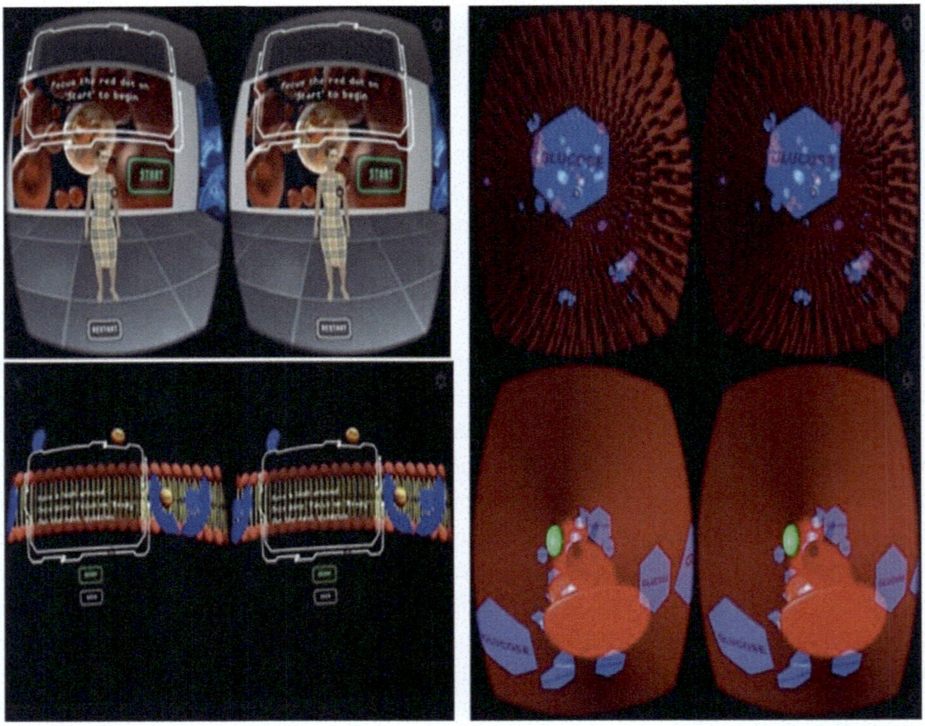

Fig. 4.6 From top left to bottom right: a classroom and the inside of the intestines, a cell membrane, and a blood vessel (Jansen et al., 2023, p. 66)

these models is an important step in stimulating students' meta-modeling knowledge for both 2D and VR models" (ibid., p. 59). Also, the students that had the prior knowledge were able to extend the content of their stored knowledge after working on the VR model. This particular finding underscores unresolved general issues in XR education research regarding the level and type of prior knowledge that are either useful or needed in order to ensure that students are able to maximize their learning in XR contexts.

Considering the elements in Table 2.1, the VR app as described seems to focus primarily on how to model the biological process itself. The synergy among the three components of the design framework is quite strong, especially since the rationale for visualizing in 3D was to foster deeper forms of reasoning and thinking about the models. It would be interesting to learn more about the role of human action and interaction in the virtual environment, including the elements of embedded and extended cognition that are supported in the VR system.

4.6 MR Artificial Intelligence (AI) System to Support Young Children's Understanding of the Physics of Stability

Yannier et al.'s (2020) experimental study investigated the impact of an MR AI learning environment system called *Intelligent Science Station* on 75 first and second-grade children's (ages 5–9 years) understanding of concepts associated with physical properties of stability (i.e., principles associated with height, wide base, symmetry, and center of mass) in the following three learning conditions: guided inquiry, explore and construct inquiry, and combination of guided and open inquiry.

Working in pairs, one group of students placed two contrasting prebuilt towers on an earthquake table and then guessed the tower that would fall first when the table shook. The MR AI environment could "observe and interpret students' actions – to accurately monitor and evaluate student predictions, experiments, and explanations and provide an intelligent guided STEM learning experience for students with varied backgrounds and support" (ibid., p. 82). The intelligent system that the authors developed also provided instructional guidance (i.e., feedback and students' progress) when the students engaged "in physical experimentation in the real world" (ibid.). Figure 4.7 shows the physical environment consisting of an earthquake table, colorful blocks of towers placed on the table, a Kinect depth camera facing the objects and endowed with a specialized built-in AI vision algorithm, a projector, and a display screen with a computer game that provided personalized audio visual feedback. Both the camera and AI algorithm could see a number of things, such as detecting and tracking movement of students' actions and physical objects on the table, monitoring how quickly a tower of objects falls down, and seeing the kinds of towers that the students constructed. Overall, the MR AI system automated some aspects of the experimental process for the students.

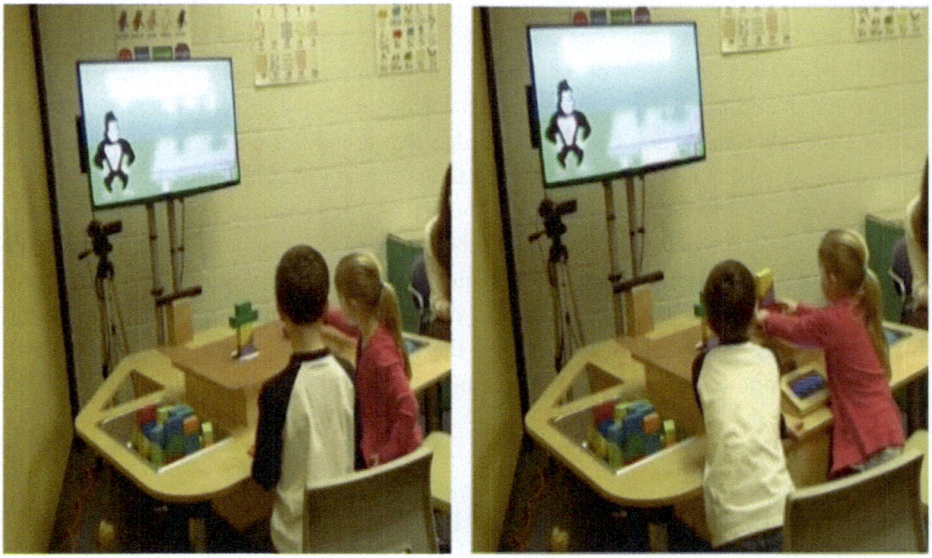

Fig. 4.7 MR AI learning environment for exploring stability (Yannier et al., 2020)

Figure 4.8 illustrates the MR AI system when it is programmed to engage students in deliberate practice or guided discovery in a highly structured inquiry setting. A gorilla character provides students with different kinds of feedback and guidance to help them learn to think critically. Toward the end of each activity, the gorilla character would then confirm whether the students' explanation was correct or incorrect "with a visualization that illustrates the corresponding physics principle" (ibid., p. 83).

Also working in pairs, another group of students engaged in a more authentic building activity that had them initially constructing different towers. The gorilla character would then ask them questions such as "Can you make a tower that will stay up when the table shakes? Place your tower on the table and click SHAKE when you are ready" (ibid., p. 84). Once they saw the result of their construction, the gorilla character would either prod them to try again if their tower fell down or congratulate them if their tower stayed up. They also engaged in different kinds of challenges that allowed them to test and solidify their understanding of physical stability principles.

A third group of students also explored the same activity, but in their case they engaged in both guided structured and open inquiry processes. All three groups of children in different learning conditions completed pre- and post- paper-based assessments, including a tower building construction activity after the post-test to measure how they incorporated their understanding of stability in their design and a survey that asked them how much they enjoyed the tower building activity.

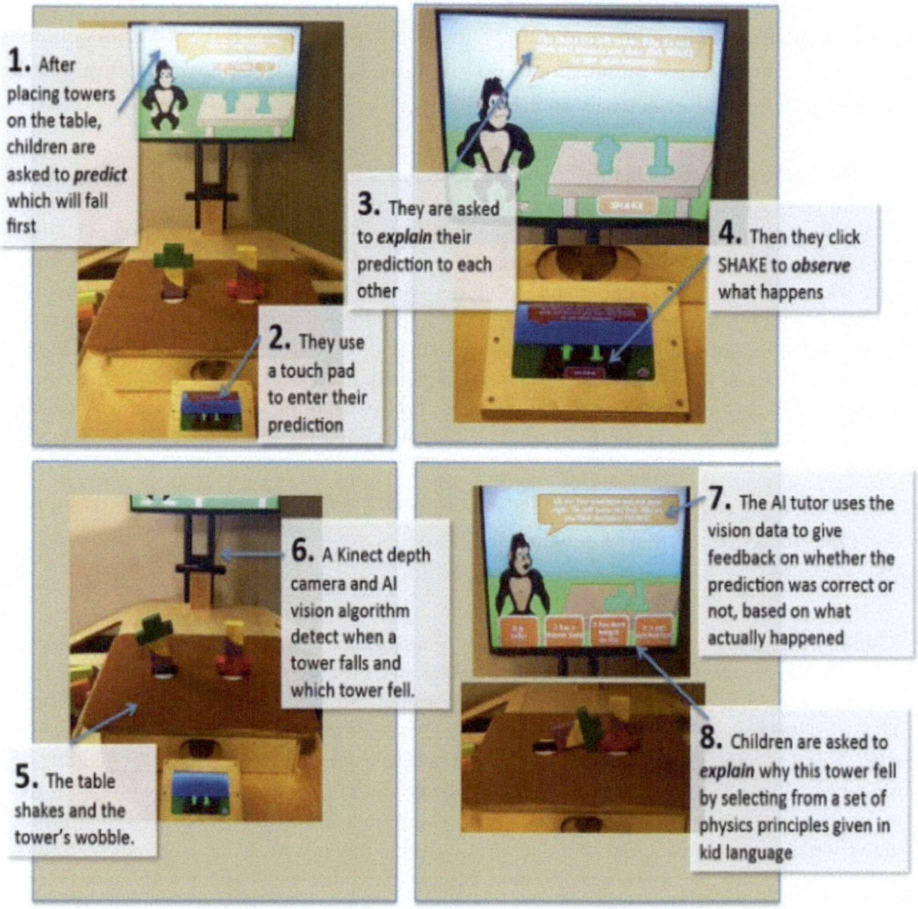

Fig. 4.8 Guided inquiry condition (Yannier et al., 2020)

The authors' findings indicate that students in the guided inquiry and combined guided and open inquiry conditions, with the help of the MR AI system, modeled better understanding, "formulated better scientific theories" and predictions, were "better able to learn to apply science in engineering tasks" (ibid., p. 91), and more generally were more transfer-appropriate ready in comparison with the students in the open inquiry condition. Their results support their suggestion regarding incorporating some "exploratory construction" to improve guided inquiry. Also, the authors suggest that "students [may] develop deeper understanding and can transfer better to real-world building if they learn some fundamentals [via guided inquiry] in service of learning higher level thinking (e.g. problem solving via Explore-Construct)" (ibid., p. 93). Regarding the students in the open inquiry condition, the authors underscored their rather "limited relevant prior knowledge" that could explain the "unsystematic and random tweaking" (e.g., "maybe I can go like this" or "you just stack them")" that they modeled, which was not well conceptualized and planned. The authors' suggestion below is quite insightful in terms of how learning might be implemented effectively in XR settings.

> We suspect the sequencing in the Combined condition, with Guided-Discovery first, is important. Initial practice on the fundamental principles, which the guided discovery provides, may aid thinking and learning during the building challenges. When children fail during the building challenges in the Combined condition, they bring to bear the scientific principles they have learned to iterate and build better structures – a kind of productive failure. The Explore-Construct condition, in contrast, produces more "unproductive failure" because children tend to respond to failure with random tweaking when they do not have more principle knowledge to employ. Without the prior preparation through guided discovery of the scientific principles, it is too complex (too much cognitive load) and/or time consuming (too big a discovery search space) for children to discover these principles through building. (Yannier et al., 2020, p. 93)

Furthermore, their recommendations below indicate the usefulness of explicitly attending to the type of learning that we want our students to model in such settings prior to implementation.

> (C)hildren learning with AI adaptive inquiry guidance in a mixed-reality setting achieve greater understanding of the scientific principles and over ten times more transfer to real world construction/engineering than children learning through hands-on construction alone. These results suggest that mixed-reality/augmented-reality systems may be more effective when forms of interactive guidance are added to facilitate active learning and understanding of fundamental scientific principles, rather than having a sole focus on hands-on exploration. To make such interactive guidance possible, may require innovation in AI tutoring support, like the vision algorithms used here, but the efforts seem to pay off. More generally, these results suggest that hands-on experiences are more effective learning experiences with more potential to improve STEM education, when integrated with guided inquiry and deliberate practice activities. (Yannier et al., 2020, p. 94)

Figure 4.9 is a revision of the design framework model shown in Fig. 2.4, which incorporates the need to attend to different conceptions of learning and pedagogy that shape the way XR apps are further designed, implemented, and experienced (cf. Garzón et al., 2020; Marougkas et al., 2023; Matovu, et al., 2023; Penn & Ramnarain, 2023; Radianti et al., 2020). For instance, the constructionist approach and direct or explicit instruction model convey two contrasting ways in which learning can take place, which are likely to affect how teachers and students use XR apps in the classroom.

Finally, in terms of the elements listed in Table 2.1, Yannier et al.'s MR AI system exemplifies what we all hope XR apps and the learning environments they support would be able to do for all learners – that is, to physically/bodily engage with concepts in either physical or digital format, generate knowledge and thinking through acting on objects, use the hybrid environment to model parity and influence their thinking, manipulate objects

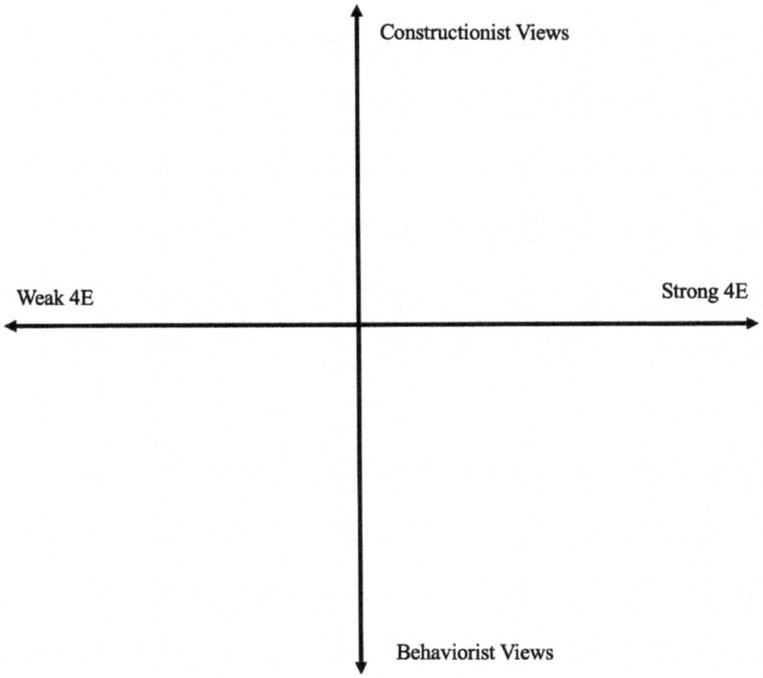

Fig. 4.9 Conception of learning × design framework for XR apps

to confirm what they know and do not know, and engage in cognitive niche constructions. The authors' design framework also reflects a strong synergy between content/curriculum/tasks (i.e., principles of stability, the construction tasks), app (i.e., the MR AI system), and formative assessment (i.e., the strong emphasis on children's reasoning).

References

Al-Ansi, A. M., Jaboob, M., Garad, A., & Al-Ansi, A. (2023). Analyzing augmented reality (AR) and virtual reality (VR) recent development in education. *Social Sciences & Humanities Open, 8*(1), 100532–100542.

Alalwan, N., Cheng, L., Al-Samarraie, H., Yousef, R., Alzahrani, A. I., & Sarsam, S. M. (2020). Challenges and prospects of virtual reality and augmented reality utilization among primary school teachers: A developing country perspective. *Studies in Educational Evaluation, 66*, 100876–100888.

Altynpara, E. & Chabanovska, D. (2023). The Fullest guide on location-based AR apps development. *Cleveroad.* Retrieved online: https://www.cleveroad.com/blog/location-based-ar-apps-development/

Antonietti, A., Rasi, C., Imperio, E., & Sacco, M. (2000). The representation of virtual reality in education. *Education and Information Technologies, 5*(4), 317–327.

Apple. (2019). *Create geometry in augmented reality.* Retrieved online at https://www.youtube.com/watch?v=e97kHReun18

Aubrey J. S., Robb M. B., Bailey J., & Bailenson J. (2018). Virtual reality 101: What you need to know about kids and VR. *Common Sense.*

Auliya, R. N., & Munasiah, M. (2019). Mathematics learning instrument using augmented reality for learning 3D geometry. *Journal of Physics. Conference Series, 1318*(1), 12069–12073.

Bailenson, J. (2018). *Experience on demand: What virtual reality is, how it works, and what it can do.* W. W. Norton & Company.

Barad, K. (2007). *Meeting the universe halfway: Quantum physics and the entanglement of matter and meaning.* Duke University Press.

Battarbee, K., & Koskinen, I. (2005). Co-experience: User experience as interaction. *CoDesign, 1*(1), 5–18.

Benedetti, P., & DeHart, N. (1997). *Forward through the rearview mirror: Reflections on and by Marshall McLuhan* (1st ed.). MIT Press.

Bujak, K. R., Radu, I., Catrambone, R., MacIntyre, B., Zheng, R., & Golubski, G. (2013). A psychological perspective on augmented reality in the mathematics classroom. *Computers and Education, 68*, 536–544.

Cabada, R. Z., López, H. M. C., & Escalante, H. J. (2023). *Multimodal affective computing: Technologies and applications in learning environments.* Springer.

Cai, S., Liu, C., Wang, T., Liu, E., & Liang, J. (2021). Effects of learning physics using augmented reality on students' self-efficacy and conceptions of learning. *British Journal of Educational Technology, 52*(1), 235–251.

F. Rivera, *STEM Learning in Extended Reality*, Synthesis Lectures on Engineering, Science, and Technology, https://doi.org/10.1007/978-3-031-78973-1

Cai, S., Liu, E., Shen, Y., Liu, C., Li, S., & Shen, Y. (2020). Probability learning in mathematics using augmented reality: Impact on student's learning gains and attitudes. *Interactive Learning Environments, 28*(5), 560–573.

Castro-Alonso, J. C., Wong, M., Adesope, O. O., Ayres, P., & Paas, F. (2019). Gender imbalance in instructional dynamic versus static visualizations: a meta-analysis. *Educational Psychology Review, 31*(2), 361–387.

Chang, H.-Y., Binali, T., Liang, J.-C., Chiou, G.-L., Cheng, K.-H., Lee, S.W.-Y., & Tsai, C.-C. (2022). Ten years of augmented reality in education: a meta-analysis of (quasi-) experimental studies to investigate the impact. *Computers and Education, 191*, 104641–104665.

Chen, J., Dai, J., Zhu, K., & Xu, L. (2022). Effects of extended reality on language learning: A meta-analysis. *Frontiers in Psychology, 13*, 1016519–1016532.

Chiang, T. H. C., Yang, S. J. H., & Hwang, G.-J. (2014). Students' online interactive patterns in augmented reality-based inquiry activities. *Computers and Education, 78*, 97–108.

Chiang, Y.-C., & Liu, S.-C. (2023). The effects of extended reality technologies in STEM education on students' learning response and performance. *Journal of Baltic Science Education, 22*(4), 568–578.

Clark, A., & Chalmers, D. (1998). The Extended Mind. *Analysis, 58*(1), 7–19.

Clark, A. (2007). Soft selves and ecological control. In D. Ross, D. Spurrett, H. Kincaid, & G. L. Stephens (Eds.), *Distributed cognition and the will* (pp. 101–122). MIT Press.

Clark, A. (2008). Supersizing the mind: Embodiment, action, and cognitive extension. Oxford University Press.

Clark, R. E. (1985). Evidence for confounding in computer-based instruction studies: Analyzing the meta-analyses. *Educational Communication and Technology, 33*(4), 249–262.

Clarke-Midura, J., & Dede, C. (2010). Assessment, technology, and change. *Journal of Research on Technology in Education, 42*(3), 309–328.

Coban, M., Bolat, Y. I., & Goksu, I. (2022). The potential of immersive virtual reality to enhance learning: A meta-analysis. *Educational Research Review, 36*, 100452–100471.

Corrêa, N. K., Galvão, C., Santos, J. W., Del Pino, C., Pinto, E. P., Barbosa, C., Massmann, D., Mambrini, R., Galvão, L., Terem, E., & de Oliveira, N. (2023). Worldwide AI ethics: A review of 200 guidelines and recommendations for AI governance. *Patterns (New York, N.Y.), 4*(10), 100857–100857.

Croom, S., Zhou, H., & Firestone, C. (2023). Seeing and understanding epistemic actions. *Proceedings of the National Academy of Sciences - PNAS, 120*(49), e2303162120–e2303162122.

Denizli-Gulboy, H., Genc-Tosun, D., & Gulboy, E. (2023). Evaluating augmented reality as evidence-based practice for individuals with autism spectrum disorder: A meta-analysis of single-case design studies. *International Journal of Developmental Disabilities, 69*(4), 472–486.

Di Paolo, E., Buhrmann, T., & Barandiaran, X. (2017). *Sensorimotor life: An enactive proposal.* Oxford University Press.

Dordio, A., Lancho, E., Merchán, M. J., & Merchán, P. (2024). Cultural heritage as a didactic resource through extended reality: A systematic review of the literature. *Multimodal Technologies and Interaction, 8*(7), 58–78.

Dudley, J., Yin, L., Garaj, V., & Kristensson, P. (2023). Inclusive immersion: A review of efforts to improve accessibility in virtual reality, augmented reality and the metaverse. *Virtual Reality, 27*(4), 2989–3020.

EchoPixel. (2021). *Echopixel's true 3D gives doctors a holographic look at your insides.* Retrieved online at https://youtu.be/oJIUAQFhbJc?si=HYnZBRJW8XIEK4YB

Engel, A. K., Friston, K. J. (Karl J.), & Kragic, D. (Eds.). (2015). The pragmatic turn: toward action-oriented views in cognitive science. The MIT Press.

Erickson, M. (2017). Virtual reality system helps surgeons, reassures patients. Stanford Medicine News Center. Retrieved online at https://med.stanford.edu/news/all-news/2017/07/virtual-reality-system-helps-surgeons-reassures-patients.html.

Faulconer, E. K., & Gruss, A. B. (2018). A review to weigh the pros and cons of online, remote, and distance science laboratory experiences. *International Review of Research in Open and Distance Learning, 19*(2), 155–168.

Ferrara, A., Rapuano, M., Ruotolo, F., Ruggiero, G., & Iachini, T. (2021). Being there in virtual reality: The sense of presence is associated with the ability to imagine body movements. In O. Gigliotta & M. Ponticorvo (Eds.), *Proceedings of the Third Symposium on Psychology-Based Technologies, Virtual Event, Naples, Italy, 4–5* Volume 3100 of *CEUR Workshop Proceedings.* Retrieved online at https://ceur-ws.org/Vol-3100/paper6.pdf

Gallagher, S. (2023). *Embodied and enactive approaches to cognition.* University of Cambridge Press.

Garzón, J., Kinshuk, Baldiris, S., Gutiérrez, J., & Pavón, J. (2020). How do pedagogical approaches affect the impact of augmented reality on education? A meta-analysis and research synthesis. *Educational Research Review, 31,* 100334–100353.

Garzón, J., & Acevedo, J. (2019). Meta-analysis of the impact of augmented Reality on students' learning gains. *Educational Research Review, 27,* 244–260.

Georgieva, M., Nelson, J., LaFosse, R., & Contis, D. (2024). *Navigating the XR educational landscape: Privacy, safety, and ethical guidelines.* Retrieved online at https://www.educause.edu/research/community/2024/navigating-the-xr-educational-landscape-privacy-safety-and-ethical-guidelines/introduction

Goldman, A., & de Vignemont, F. (2009). Is social cognition embodied? *Trends in Cognitive Sciences, 13*(4), 154–159.

Gu, P., Xu, X., Qian, X., & Weng, T.-H. (2023). Leveraging extended reality for autistic individuals: A scoping review of technical features and technology affordances. *IEEE Transactions on Learning Technologies, 16*(1), 133–149.

Habig, S. (2020). Who can benefit from augmented reality in chemistry? Sex differences in solving stereochemistry problems using augmented reality. *British Journal of Educational Technology, 51*(3), 629–644.

Herold, B. (2019). What is personalized learning? Education Week. Retrieved online at https://www.edweek.org/technology/what-is-personalized-learning/2019/11

Herrera, L. M. M., Abalo, M. A., & Ordóñez, S. J. (2019). Learning Calculus with Augmented Reality and Virtual Environments. In *Proceedings of the 2019 11th International Conference on Education Technology and Computers* (pp. 18–22).

Hite, R., Childers, G., Jones, G., Corin, E., & Pereyra, M. (2021). Describing the experiences of students with ADHD learning science content with emerging technologies. *Journal of Science Education for Students with Disabilities, 24*(1), 1–34.

Hite, R. (2021). *Augmented and virtual reality in education: an examination of the impact on learning.* Retrieved online at https://cdn.zspace.com/collateral/case-studies/Augmented_and_Virtual_Reality_in_Education-An_Examination_of_the_Impact_on_Learning.pdf

Hohol, M., Wołoszyn, K., & Brożek, B. (2021). Making cognitive niches explicit: On the importance of external cognitive representations in accounting for cumulative culture. *Frontiers in Integrative Neuroscience, 15,* 734930–734930.

Innocente, C., Ulrich, L., Moos, S., & Vezzetti, E. (2023). A framework study on the use of immersive XR technologies in the cultural heritage domain. *Journal of Cultural Heritage, 62,* 268–283.

Interaction Design Foundation - IxDF. (2016). *What is virtual reality?* Retrieved online at https://www.interaction-design.org/literature/topics/virtual-reality

Jaber, O., Bagossi, S., Fried, M. N., & Swidan, O. (2024). Conceptualizing functional relationships in an augmented reality environment: Connecting real and virtual worlds. *ZDM Mathematics Education, 56*(4), 605–623.

Jansen, S., Taib, S. F. B. M., Cai, Y., & van Joolingen, W. R. (2023). Applying prior meta-modeling knowledge to a VR model of a biological process. *Mixed Reality for Education* (pp. 59–83). Springer.

Jiang, Y., Clarke-Midura, J., Baker, R. S., Paquette, L., & Keller, B. (2018). How immersive virtual environments foster self-regulated learning. In R. Zheng (Ed.), *Digital technologies and instructional design for personalized learning* (pp. 28–54). IGI Global Scientific Publishing.

Jochecová, K., Černý, M., Stachoň, Z., Švedová, H., Káčová, N., Chmelík, J., Brůža, V., Kvarda, O., Ugwitz, P., Šašinková, A., Fořtová, N. C., & Šašinka, Č. (2022). Geography education in a collaborative virtual environment: A qualitative study on geography teachers. *ISPRS International Journal of Geo-Information, 11*(3), 180–195.

Johnson, M. (2017). *Embodied mind, meaning, and reason: How our bodies give rise to understanding*. University of Chicago Press.

Kang, K., Kushnarev, S., Wei Pin, W., Ortiz, O., & Chen Shihang, J. (2020). Impact of virtual reality on the visualization of partial derivatives in a multivariable calculus class. *IEEE Access, 8*, 58940–58947.

Kaplan, A. D., Cruit, J., Endsley, M., Beers, S. M., Sawyer, B. D., & Hancock, P. A. (2021). The effects of virtual reality, augmented reality, and mixed reality as training enhancement methods: a meta-analysis. *Human Factors, 63*(4), 706–726.

Kaufmann, H. (2009). Virtual environments for mathematics and geometry education. *Themes in Science and Technology Education, 2*(1–2), 131–152.

Keen, A. (2018). Book world: Can VIRTUAL REALITY MAKE EVERYTHING BETTER?(Experience on Demand: What Virtual Reality Is, How It Works, and What It Can Do) [Review of *Book World: Can virtual reality make everything better?(Experience on Demand: What Virtual Reality Is, How It Works, and What It Can Do)*]. *The Washington Post*. The Washington Post.

Kelly, J., Gilbert, S., Dorneich, M., & Costabile, K. (2023). Gender differences in cybersickness: Clarifying confusion and identifying paths forward. In: *IEEE Conference on Virtual Reality and 3D User Interfaces Abstracts and Workshops* (pp. 283–288)

Kerawalla, L., Luckin, R., Seljeflot, S., & Woolard, A. (2006). Making it real : Exploring the potential of augmented reality for teaching primary school science. *Virtual Reality, 10*(3–4), 163–174.

Khaddage, F., Lattemann, C., & Gebbing, P. (2023). Towards User2Machine model for higher education—enforced by Covid19 pandemic. In M. E. Auer et al. (Eds.), *Learning in the age of digital and green transition* (pp. 163–171). Springer.

Khan, M., Trujano, F., Choudhury, A., & Maes, P. (2018). Mathland: playful mathematical learning in mixed reality. In: *Extended Abstracts of the 2018 CHI Conference on Human Factors in Computing Systems* (pp. 1–4).

Kiourt, C., Kalles, D., Lalos, A., Papastamatiou, N., Silitziris, P., Paxinou, E., Theodoropoulou, H., Zafeiropoulos, V., Papadopoulos, A., & Pavlidis, G. (2020). XRLabs: Extended reality interactive laboratories. In: *Proceedings of the 12th International Conference on Computer Supported Education—Volume 1: CSEDU* (pp. 601–608).

Kirsh, D., & Maglio, P. (1994). On distinguishing epistemic from pragmatic action. *Cognitive Science, 18*(4), 513–549.

Korkmaz, E., & Morali, H. S. (2022). A meta-synthesis of studies on the use of augmented reality in mathematics education. *International Electronic Journal of Mathematics Education, 17*(4), 1–21.

Lakoff and Johnson (1991)

Lakoff, G., & Johnson, M. (1999). *Philosophy in the flesh: The embodied mind and its challenge to western thought*. New York: Basic Books.

Magnani, L. (2007). Creating chances through cognitive Niche construction. In B. Apolloni, R. J. Howlett, & L. Jain (Eds.), *Knowledge-based intelligent information and engineering systems*. KES 2007. Lecture Notes in Computer Science (Vol. 4693, pp. 917–925). Springer.

Mann, S., Furness, T., Yu, Y., Iorio, J., & Wang, Z. (2018). All Reality: Virtual, Augmented, Mixed (X), Mediated (X,Y), and Multi-mediated Reality. *arXiv (Cornell University)*. Retrieved online at https://doi.org/10.48550/arXiv.1804.08386

Marougkas, A., Troussas, C., Krouska, A., & Sgouropoulou, C. (2023). Virtual reality in education: A review of learning theories, approaches and methodologies for the last decade. *Electronics (Basel), 12*(13), 2832–2853.

Massaro, D. (2019). Experiencing reality in virtual reality [Review of Experiencing Reality in Virtual Reality]. *American Journal of Psychology, 132*(4), 495–500.

Matovu, H., Ungu, D. A. K., Won, M., Tsai, C.-C., Treagust, D. F., Mocerino, M., & Tasker, R. (2023). Immersive virtual reality for science learning: Design, implementation, and evaluation. *Studies in Science Education, 59*(2), 205–244.

Mayer, R. E., Makransky, G., & Parong, J. (2023). The promise and pitfalls of learning in immersive virtual reality. *International Journal of Human-Computer Interaction, 39*(11), 2229–2238.

Mecharithm. (2022). *Augmented reality (AR) scene with 2D image marker and ground detection*. Retrieved online at https://youtu.be/GJYb5gbDgIY?si=7qYw9D4AosA-b7aJ

Mecharithm. (2021). *Location-based Augmented Reality (AR) Using GPS*. Retrieved online at https://www.youtube.com/watch?v=9ltZuIJKLOI

Menary, R. (2007). Cognitive Integration: Mind and Cognition Unbounded. New York: Palgrave-Macmillan.

Menary, R. (2010). Introduction to the special issue on 4E cognition. *Phenomenology and the Cognitive Sciences, 9*(4), 459–463.

Miłkowski, M., Clowes, R., Rucińska, Z., Przegalińska, A., Zawidzki, T., Krueger, J., Gies, A., McGann, M., Afeltowicz, Ł, Wachowski, W., Stjernberg, F., Loughlin, V., & Hohol, M. (2018). From wide cognition to mechanisms: A silent revolution. *Frontiers in Psychology, 9*, 2393–2410.

Mislevy, R. J., & Haertel, G. D. (2006). Implications of evidence-centered design for educational testing. *Educational Measurement, Issues and Practice, 25*(4), 6–20.

Mosher, M. A., Carreon, A. C., Craig, S. L., & Ruhter, L. C. (2022). Immersive technology to teach social skills to students with autism spectrum disorder: A literature review. *Review Journal of Autism and Developmental Disorders, 9*(3), 334–350.

Di Natale, A. F., Repetto, C., Riva, G., & Villani, D. (2020). Immersive virtual reality in K-12 and higher education: A 10-year systematic review of empirical research. *British Journal of Educational Technology, 51*(6), 2006–2033.

Newen, A., De Bruin, L., & Gallagher, S. (2018). *The Oxford handbook 4E cognition*. Oxford University Press.

Newton, M. H., Annetta, L. A., & Bressler, D. M. (2024). Using extended reality technology in traditional and place-based environments to study climate change. *Journal of Science Education and Technology, 33*(2), 208–227.

Nguyen, H., & Bednarz, T. (2020). User experience in collaborative extended reality: Overview study. In *17th EuroVR International Conference, Valencia, Spain* (pp. 41–70). Retrieved online at https://hal.science/hal-03030601/document

Pahmi, S., Hendriyanto, A., Sahara, S., Muhaimin, L. H., Kuncoro, K. S., & Usodo, B. (2023). Assessing the Influence of augmented reality in mathematics education: A systematic literature review. *International Journal of Learning, Teaching and Educational Research, 22*(5), 1–25.

Pang, G. C., & Cai, Y. (2023). Transforming learning experiences through affordances of virtual and augmented reality. In H. B.-L. Duh (Ed.), *Mixed reality for education: Gaming media and social effects* (pp. 109–165). Springer.

Penn, M., & Ramnarain, U. (2023). A systematic review of pedagogy related to mixed reality in K-12 education. *Mixed Reality for Education* (pp. 85–108). Springer.

Petrov, P. D., & Atanasova, T. V. (2020). The effect of augmented reality on students' learning performance in STEM education. *Information, 11*(4), 209–220.

Pinker, S. (2010). Cognitive niche: Coevolution of intelligence, sociality, and language. *Proceedings of the National Academy of Sciences—PNAS, 107*(Supplement 2), 8993–8999.

PlugXR. (2024). *Location-based AR in 2024—A complete guide.* Retrieved online at https://www.plugxr.com/augmented-reality/location-based-ar/

Potkonjak, V., Gardner, M., Callaghan, V., Mattila, P., Guetl, C., Petrović, V. M., & Jovanović, K. (2016). Virtual laboratories for education in science, technology, and engineering: A review. *Computers and Education, 95*, 309–327.

Radianti, J., Majchrzak, T. A., Fromm, J., & Wohlgenannt, I. (2020). A systematic review of immersive virtual reality applications for higher education: Design elements, lessons learned, and research agenda. *Computers and Education, 147*, 103778–1037807.

Rahimi, S., & Shute, V. J. (2023). Stealth assessment: a theoretically grounded and psychometrically sound method to assess, support, and investigate learning in technology-rich environments. *Educational Technology Research and Development*, 1–25.

REL West. (2021*). Effect size basics: Understanding the strength of a program's impact.* Retrieved online at https://ies.ed.gov/ncee/edlabs/regions/west/relwestFiles/pdf/4-2-3-14_Effect_Size_Infographic_Final_508c.pdf

Rivera, F., & Paidi, C. (2024). *VR transforming graphs app: Development and design process.* Technical Report, San Jose State University.

Rivera, F., Koev, P., Liu, Y., & Stafford, P. (2023). *VR linear algebra: Development and design process.* Technical Report, San Jose State University.

Rogers, S. (2019). *Virtual reality: The learning aid of the 21st century.* Retrieved online at https://www.forbes.com/sites/solrogers/2019/03/15/virtual-reality-the-learning-aid-of-the-21st-century/

Schaper, M.-M., & Pares, N. (2022). Enhancing students' social and emotional learning in educational virtual heritage through projective augmented reality. In *CHI Conference on Human Factors in Computing Systems Extended Abstracts* (pp. 1–9).

Shi, Z., Bonar, J., Wang, Y., & Silverstein, B. (2024). Navigating the Future with VR/AR/MR technologies: A perspective from industry leaders. *IEEE Journal of Selected Topics in Quantum Electronics, 30*(2), 1–4.

Shute, V. J., & Rahimi, S. (2022). Stealth assessment: A Primer. *Rapid Community Report Series.* Digital Promise and the International Society of the Learning Sciences. https://repository.isls.org//handle/1/7671, pp. 1–11.

Shute, V. J. (2023). History of stealth assessment and a peek into its future. In M. P. McCreery & S. K., Krach (Eds.), *Games as stealth assessments* (pp. 1–23). IGI Global. https://doi.org/10.4018/979-8-3693-0568-3.ch00110.4018/979-8-3693-0568-3.ch001

Simeone, A. (Ed.). (2023). *Everyday virtual and augmented reality.* Springer.

Slater, M., & Sanchez-Vives, M. V. (2016). Enhancing our lives with immersive virtual reality. *Frontiers in Robotics and AI, 3*, 1–47.

Smart, P. R., & Sycara, K. (2015). Situating cognition within the virtual world. *Procedia Manufacturing, 3*, 3836–3843.

Smith, G., Shute, V. J., Rahimi, S., Kuba, R., & Dai, C.-P. (2023). Stealth assessment and digital learning game design. In M. P. McCreery, & S. K., Krach (Eds.), *Games as stealth assessments* (pp. 81–100). IGI Global.

Sommerauer, P., & Müller, O. (2014). Augmented reality in informal learning environments: A field experiment in a mathematics exhibition. *Computers and Education, 79*, 59–68.

Song, E., Suaib, N. M., Sihes, A. J., Alwee, R., & Yunos, Z. M. (2020). Design and development of learning mathematics game for primary school using handheld augmented reality. *IOP Conference Series. Materials Science and Engineering, 979*(1), 12014–12024.

Sprevak, M. (2019). Extended cognition. *Routledge Encyclopedia of Philosophy*. Retrieved online at https://www.rep.routledge.com/articles/thematic/extended-cognition/v-1

Sutton, J. (2010). Exograms and interdisciplinarity: History, the extended mind, and the civilising process. In R. Menary (Ed.), *The extended mind* (pp. 189–226). Cambridge, MA: The MIT Press.

Tekedere, H., & Göke, H. (2016). Examining the effectiveness of augmented reality applications in education: A meta-analysis. *International Journal of Environmental and Science Education, 11*(16), 9469–9481.

Tene, T., Guevara, M., Moreano, G., Vera, J., & Vacacela Gomez, C. (2024). The role of immersive virtual realities: Enhancing science learning in higher education. *Emerging Science Journal, 8*, 88–102.

Thelen, E., & Bates, E. (2003). Connectionism and dynamic systems: Are they really different? *Developmental Science, 6*, 378–391.

Tirado, C., Khatin-Zadeh, O., Gastelum, M., Leigh-Jones, N., & Marmolejo-Ramos, F. (2018). The strength of weak embodiment. *International Journal of Psychological Research, 11*(2), 77–85.

Tooby, J., & DeVore, I. (1987). The reconstruction of hominid evolution through strategic modeling. In W. G. Kinzey (Ed.), *The evolution of human behavior: Primate models* (pp. 183–237). SUNY Press.

Tremosa, L (2022). *Virtuality continuum*. Retrieved online at https://www.interaction-design.org/literature/topics/virtuality-continuum?srsltid=AfmBOooc3NoiE9LN27iAAB_vzCH9K_k0NjIWvDBZ70kxG3CuJnWtfA9P

Tzima, S., Styliaras, G., & Bassounas, A. (2019). Augmented reality applications in education: Teachers point of view. *Education Sciences, 9*(2), 99–117.

Ulbrich, E., Haas, B., Elbedewy, S., & Lavicza, Z. (2023). Amazing walk through mathematics. *Mixed reality for education* (pp. 7–31). Springer.

Wikitude. (2024). *Geo AR: Location-based augmented reality use cases and how-to*. Retrieved online at https://www.wikitude.com/blog-geo-ar-location-based-augmented-reality/

Yannier, N., Hudson, S. E., & Koedinger, K. R. (2020). Active Learning is about more than hands-on: A mixed-reality AI system to support STEM education. *International Journal of Artificial Intelligence in Education, 30*(1), 74–96.

Yilmaz, Z. A., & Batdi, V. (2021). Meta-analysis of the use of augmented reality applications in science teaching. *Journal of Science Learning (Online), 4*(3), 267–274.

Ziker, C., Truman, B., & Dodds, H. (2021). Cross reality (XR): Challenges and opportunities across the spectrum. In J. Ryoo & K. Winkelmann (Eds.), *Innovative learning environments in STEM higher education* (pp. 55–77). Springer.

zSpace. (2023). Innovative AR/VR Technology. Retrieved online at https://zspace.com/technology.

The manufacturer's authorised representative in the EU is Springer Nature Customer Service Centre GmbH, Europaplatz 3, 69115 Heidelberg, Germany. If you have any concerns regarding our products, please contact ProductSafety@springernature.com

Printed and bound by CPI Group (UK) Ltd, Croydon, CR0 4YY
01/10/2026
02225980-0019